For information on translations, please email *bmharwani@yahoo.com*

Edition: 2010

Dedicated to all my school friends (St. Anselm's School, Ajmer, India) and Engineering College friends (S.S.V.P.S's BapuSaheb ShivajiRao Deore College of Engineering, Maharashtra, India)

Acknowledgements

I owe a debt of gratitude to my friend and project editor, Susan Glinert Stevens, for her cooperation and first-class structural and language editing. I appreciate her efforts in enhancing the contents and giving the book a polished appearance.

I am also thankful to my family—my small world: Anushka (my wife) and my two little darlings, Chirag and Naman, for allowing me to work on the book when I was supposed to spend time with them.

I should not forget to thank my dear students, who have been good teachers for me, as they make me aware of the basic programming problems they face, enabling me to address that which puzzles them. The endless, interesting student queries has helped me to write books with a practical approach.

About the Author

 B. M. Harwani is the managing director of the Computer Education Centre - Microchip Computer Education (MCE), based in Ajmer, India. He graduated with a BE in computer engineering from the University of Pune, and has a 'C' Level master's diploma in computer technology from DOEACC, Government Of India. Involved in teaching field for over 15 years, Mr. Harwani has developed the art of explaining even the most complicated topics in a manner that everybody can easily understand. He has written many successful books, including *Programming & Problem Solving through C* (BPB, 2004), *Learn Tally in Just Three Weeks* (Pragya, 2005), *Data Structures and Algorithms through C* (CBC, 2006), *Master Unix Shell Programming* (CBC, 2006), *Business Systems* (CBC, 2006), *Practical Java Projects* (Shroff, 2007), *Practical Web Services* (Shroff, 2007), *Java for Professionals* (Shroff, 2008), *C++ for Beginners* (Shroff, 2009), *Practical ASP.NET 3.5 Projects* (Shroff, 2009), *Java Server Faces—A Practical Approach for Beginners* (PHI Learning, 2009), *Practical JSF Project using NetBeans* (PHI Learning, 2009), *Foundation Joomla* (Friends of ED, 2009), *Practical EJB Projects* (Shroff, 2009), *Data Structures and Algorithms in C++* (Dreamtech Press, 2010), *Developing Web Applications in PHP and AJAX* (Tata McGraw Hill, 2010), and *jQuery Recipes* (Apress, 2010). He can be contacted at **bmharwani@yahoo.com**.

Table of Contents

1.

Introduction

Welcome to *Beginning Web Development for Smartphones*! The past few years have seen a meteoric rise in mobile devices, so it's not surprising that developers are scrambling to create programs adapted to the small screens of the iPhone, Blackberry, and Android devices. This book will help you get started—demystifying the components, processes, and code required to write your own applications.

Once you understand the basics, you'll be ready to address the vast audience of mobile users with applications for driving directions, public transportation, news, social networking, banking, book stores, maps, weather, and shopping.

The sample web application we will be developing in this book makes use of HTML, PHP, MySQL, Apache server, and jQTouch. In addition, I assume that the reader understands HTML, CSS, and JavaScript (preferably jQuery).

1.1 The Sample Application

The sample web application we will be working on is a *Book Store* application. The web application will display book categories and subcategories that the user can browse and then select items for purchase. Our web application will have the following characteristics :

- The book categories and subcategories shown to the user will not be hard coded; instead, they will be fetched from server-side database tables. This means we can easily add or delete categories and subcategories without altering the actual code.

- Users can create accounts with authentication.

- Users can see detailed information for a selected book.

- Users can add books to a shopping cart, delete them, or change the order quantity.

I assume that you have installed and configured Apache, PHP, and MySQL on your machine. If not, refer the Appendix A for installation instructions.

1.2 Introduction to Required Components

The three most important components required in developing and testing a dynamic web application are: web server, scripting language, and a database. For our application, I have chosen Apache as a web server. PHP as the scripting language, and MySQL as the database.

Before we proceed further, we're going to briefly review these components for those who might them unfamiliar. If you are already an expert, feel free to skip down to Section 1.3.

1.2.1 Apache

A *web server* is a program that is loaded and run on a server computer—the machine that hosts web applications. The most popular web server apps are: Microsoft's *Internet Information Server (IIS)*and Apache Server Foundation's *Apache HTTP Server*. Web application are accessed by the client via a browser. The browser requests the web application from the server, which then executes the program. The main role of the web server is to receive the requests from the clients; execute the relevant scripts to process information passed to it; or retrieve required information from server-side databases and send the response back to the client in the form of an HTML document. This entire process is also known as a *web request life cycle*.

Note: In this book, the term *client* means any mobile device such as the iPhone, Blackberry, or Android machine.

Web Request Life Cycle

For Apache, PHP, and MySQL, the web request life cycle is as follows:

1. The client makes a request to an Apache web server in the form of an HTTP GET or POST message.

2. The Apache web server parses the request, and, depending on the request, locates the desired PHP script and executes it.

3. The PHP script either fetches the desired information from the MySQL database or updates its contents.

4. The MySQL database returns the desired information and the database status to the PHP script.

5. The PHP script combines the database information with an HTML template and sends it to the Apache web server program.

6. Apache sends an HTTP response in the form of an HTML document to the client's browser.

1.2.2 PHP

PHP stands for *Hypertext Preprocessor* and is a very popular web scripting engine. (The acronym is derived from the original name, Personal Home Page). It's the first choice of web developers for creating dynamic web pages because it's extremely flexible and very easy to learn. Being a server-side scripting language, all PHP scripts are executed on the server and the output is sent to the client as plain HTML. As a result, if anybody tries to snoop inside our website's source code, only the HTML code generated by the execution of the PHP script will be visible. The PHP script itself cannot be viewed, making it secure.

A few of PHP's features are:

- It supports a variety of databases (MySQL, Informix, Oracle, Sybase, Solid, PostgreSQL, Generic ODBC, and so on), so data can be stored or retrieved from the back-end database.

- It's open source software (OSS) freely available on the Web, which reduces software development costs. Open source also ensures faster bug fixing and quick integration of new and enhanced features.

- It can be easily embedded with HTML tags and scripts.

- It runs on different platforms, such as Windows, Linux, and Unix.

- It is compatible with almost all servers used today, for example, Apache and IIS.

- It is a server-side scripting language and is interpreted by the web server. PHP returns simple HTML code to the client, thus reducing the load on the end-user's device.

We will learn more about PHP programming in later chapters.

1.2.3 MySQL

MySQL is one of the most popular relational database management system in use today. It's open source software released under the GNU General Public License (GPL) and is fast, reliable, and very easy to learn. Above all, it's free for most uses on all supported platforms.

The benefits of storing information in databases are many. Fetching data is much faster than from traditional file systems, as databases use indexing, hashing, and other schemes to quickly find the desired data. Databases usually have auto-backup and restore facilities, encryption for high security, and built-in integrity constraints.

We won't be using many SQL commands for developing our web application, but you can refer Appendix B for the basics of MySQL.

1.3 Designing the Database

As we mentioned above, we'll be developing a sample web application called Book Store. We'll need four database tables to store the necessary information. So the first step is to open MySQL and create a database containing four tables.

Let's name the database s*hopping* and the database tables *books, customers, orders* and *orders_details.*

- The *books* table will be used to store publication information such as title, author, categories and subcategories, publisher, and publication date.

- The *customers* table will be used to store the name, address, contact number, and email address of customers purchasing the books.

- The *orders* table will be used to store the order number, order date, and information about the customer placing the order.

- The *orders_details* table will be used to store order details, such as titles, quantities, and prices of all the books contained in a specific order number.

We will not be using any server-side database tables in our application. Instead, we'll use a client-side database to store information about the user's shopping cart contents. We'll learn how to work with local databases in later chapters.

You guys won't have to create database or the tables by hand. Just execute the SQL script in Listing 1.1—the database and tables will be created automatically.

Listing 1.1. SQL Script for Creating the Database and Tables

```
create database shopping;
use shopping;


create table books (
isbn varchar(30) not null,
title varchar(100) not null,
author1 varchar(50) not null,
author2 varchar(50),
author3 varchar(50),
category varchar(50),
subcategory varchar(50),
quantity smallint,
publisher varchar(100),
publish_date_edition  varchar(50),
price float,
image varchar(50),
```

```
description text,
primary key(isbn));

create table customers (
userid varchar(50) not null,
password varchar(50),
name varchar(50),
address varchar(200),
city varchar(50),
state varchar(50),
zipcode varchar(12),
emailid varchar(50),
contact_no varchar(50),
country varchar(50),
primary key(userid));

create table orders (
order_no integer not null auto_increment,
order_date date,
userid varchar(50),
shipping_address varchar(200),
shipping_city varchar(50),
shipping_state varchar(50),
shipping_country varchar(50),
shipping_zipcode varchar(15),
credit_card_name varchar(30),
credit_card_number varchar(15),
card_expirydate date,
primary key (order_no));

create table orders_details (
order_no integer not null,
isbn varchar(30) not null,
title varchar(100) not null,
quantity integer not null,
price float
);
```

The first line of the script creates a database called *shopping*, which is loaded and activated. Four *create table* commands then generate four database tables called *books, customers, orders,* and *orders_details*.

1.3.1 The books Table

The *books* table stores the following book information: isbn number, title, author(s), categories, and subcategories, publisher, publication date, price, image, and description. The structure of the table is shown in Figure 1.1. The field contents are further explained in Table 1.1.

```
mysql> describe books;
+----------------------+--------------+------+-----+---------+-------+
| Field                | Type         | Null | Key | Default | Extra |
+----------------------+--------------+------+-----+---------+-------+
| isbn                 | varchar(30)  | NO   | PRI | NULL    |       |
| title                | varchar(100) | NO   |     | NULL    |       |
| author1              | varchar(50)  | NO   |     | NULL    |       |
| author2              | varchar(50)  | YES  |     | NULL    |       |
| author3              | varchar(50)  | YES  |     | NULL    |       |
| category             | varchar(50)  | YES  |     | NULL    |       |
| subcategory          | varchar(50)  | YES  |     | NULL    |       |
| quantity             | smallint(6)  | YES  |     | NULL    |       |
| publisher            | varchar(100) | YES  |     | NULL    |       |
| publish_date_edition | varchar(50)  | YES  |     | NULL    |       |
| price                | float        | YES  |     | NULL    |       |
| image                | varchar(50)  | YES  |     | NULL    |       |
| description          | text         | YES  |     | NULL    |       |
+----------------------+--------------+------+-----+---------+-------+
```

Figure 1.1 Structure of the *books* table

Table 1.1. The *books* Table Fields

Fields	Usage
isbn	Stores the unique isbn book code. It is a primary key and cannot be *null* for any book.
Title	Store the book title
author1,author2, author3	Stores up to three authors. If there are fewer than three, extra fields are set to *null.*
category	Stores the book category. Books can be divided among various categories such as *Literature & Fiction, Home & Garden, Computers & Internet, Entertainment*, and so on. We can add or update categories by adding or updating rows in this field without modifying the application code.
subcategory	Stores the book subcategory. For example, *Literature & Fiction* can have several subcategories such as *Drama, Novel, Short Stories, or Poetry*. Again, we can add or update subcategories of books as needed by adding or updating rows in this field.
quantity	Stores the current quantity of books in inventory.
publisher	Stores the name of the publishing house.
publish_date_edition	Stores the publication date and edition number.
price	Stores the retail price of the book.
image	Stores the folder path of a cover-page image.
description	Store a brief description of the book.

1.3.2 The customers Table

The *customers* table stores the user id, password, name, address, email, and phone number. This data can be retrieved for authentication purposes, fetching shipping information, and informing customers about new arrivals, discounts, and other information of interest. The structure of the customers table is shown in Figure 1.2. The fields are described in Table 1.2.

```
mysql> describe customers;
+-------------+--------------+------+-----+---------+-------+
| Field       | Type         | Null | Key | Default | Extra |
+-------------+--------------+------+-----+---------+-------+
| userid      | varchar(50)  | NO   | PRI | NULL    |       |
| password    | varchar(50)  | YES  |     | NULL    |       |
| name        | varchar(50)  | YES  |     | NULL    |       |
| address     | varchar(200) | YES  |     | NULL    |       |
| city        | varchar(50)  | YES  |     | NULL    |       |
| state       | varchar(50)  | YES  |     | NULL    |       |
| zipcode     | varchar(12)  | YES  |     | NULL    |       |
| emailid     | varchar(50)  | YES  |     | NULL    |       |
| contact_no  | varchar(50)  | YES  |     | NULL    |       |
| country     | varchar(50)  | YES  |     | NULL    |       |
+-------------+--------------+------+-----+---------+-------+
```

Figure 1.2 Structure of the *customers* table

Table 1.2 The *customers* Table Fields

Fields	Usage
userid	Stores the user's unique id. it is a primary key by which the customer record can be retrieved.
password	Stores a unique code for authenticating the user.
name	Stores the customer name. This field can have the same value for multiple customers, because no two customers can have the same userid.
address	Stores the customer's street address.
city	Stores the customer's city.
state	Stores the customer's state.
zipcode	Stores the customer's zip code.
emailid	Stores the customer's email address, which can be used for order confirmation and other communications.
contact_no	Stores the customer's telephone number.
country	Stores the customer's country.

1.3.3 The orders Table

The orders table stores the order number, date, user id, shipping address, and credit card information. The table is shown in Figure 1.3. The field contents are further explained in Table 1.3.

```
mysql> describe orders;
+------------------+--------------+------+-----+---------+----------------+
| Field            | Type         | Null | Key | Default | Extra          |
+------------------+--------------+------+-----+---------+----------------+
| order_no         | int(11)      | NO   | PRI | NULL    | auto_increment |
| order_date       | date         | YES  |     | NULL    |                |
| userid           | varchar(50)  | YES  |     | NULL    |                |
| shipping_address | varchar(200) | YES  |     | NULL    |                |
| shipping_city    | varchar(50)  | YES  |     | NULL    |                |
| shipping_state   | varchar(50)  | YES  |     | NULL    |                |
| shipping_country | varchar(50)  | YES  |     | NULL    |                |
| shipping_zipcode | varchar(15)  | YES  |     | NULL    |                |
| credit_card_name | varchar(30)  | YES  |     | NULL    |                |
| credit_card_number | varchar(15) | YES |    | NULL    |                |
| card_expirydate  | date         | YES  |     | NULL    |                |
+------------------+--------------+------+-----+---------+----------------+
```

Figure 1.3 Structure of the *orders* table

Table 1.3 The *orders* Table Fields

Fields	Usage
order_no	Stores a unique integer that is automatically incremented (starting with 1) and may be used by the customer for future reference. This order number is referenced in the *orders_details* table to retrieve a complete list of the books in this order.
order_date	Stores the order date.
userid	Stores the customer's unique id.
shipping_address	Stores the customer's shipping street address.
shipping_city	Stores the customer's shipping city.
shipping_state	Stores the customer's shipping state.
shipping_country	Stores the customer's shipping country.
shipping_zipcode	Stores the customer's shipping zip code.
credit_card_name	Stores the customer's credit card name and type, for example, Visa or Master Card.
credit_card_number	Stores the customer's credit card number.
card_expirydate	Stores the customer's credit card expiration date.

1.3.4 The orders_details Table

This table stores information about the books purchased on each order number. If we want a list of titles purchased in a given order, we first query the *Orders* table with the given order number and then query the *orders_details* table for that order number. The table is shown in Figure 1.4. The field contents are further explained in Table 1.4.

```
mysql> describe orders_details;
+----------+--------------+------+-----+---------+-------+
| Field    | Type         | Null | Key | Default | Extra |
+----------+--------------+------+-----+---------+-------+
| order_no | int(11)      | NO   |     | NULL    |       |
| isbn     | varchar(30)  | NO   |     | NULL    |       |
| title    | varchar(100) | NO   |     | NULL    |       |
| quantity | int(11)      | NO   |     | NULL    |       |
| price    | float        | YES  |     | NULL    |       |
+----------+--------------+------+-----+---------+-------+
```

Figure 1.4 Structure of the *orders_details* table

Table 1.4 The *orders_details* Table Fields

Fields	Usage
order_no	Stores the order number's unique integer The *order_no* from the *orders* table is automatically copied into this field.
isbn	Stores the book's unique ISBN code, which is copied from the shopping cart via the *books* table.
title	Stores the book title.
quantity	Stores the quantity purchased.
price	Stores the retail price.

1.3.5. Sample Database Listing

Figure 1.5 shows sample output with the dummy records we inserted into the *books* table. The SQL script for inserting dummy records is given in Appendix C.

```
mysql> select isbn, title, author1,category, subcategory from books;
+-------------------+-----------------------------------------+-------------+----------------------+------------------------------+
| isbn              | title                                   | author1     | category             | subcategory                  |
+-------------------+-----------------------------------------+-------------+----------------------+------------------------------+
| 111-1-1111-1111-0 | Red Queen                               | B.M.Harwani | Literature & Fiction | Drama                        |
| 111-1-1111-1111-1 | Last Block                              | B.M.Harwani | Literature & Fiction | Drama                        |
| 111-1-1111-1111-2 | All New Tales                           | B.M.Harwani | Literature & Fiction | Drama                        |
| 111-1-1111-1111-3 | The Fixer                               | B.M.Harwani | Literature & Fiction | Essays                       |
| 111-1-1111-1111-4 | Girl with Tattoo                        | B.M.Harwani | Literature & Fiction | Essays                       |
| 111-1-1111-1111-5 | Girl Who Kicked                         | B.M.Harwani | Literature & Fiction | Letters & Correspondence     |
| 111-1-1111-1111-6 | Boy Who Set Fire                        | B.M.Harwani | Literature & Fiction | Poety                        |
| 111-1-1111-1111-7 | The Dawn                                | B.M.Harwani | Literature & Fiction | Womens Fiction               |
| 111-1-1111-1111-8 | Two Sisters                             | B.M.Harwani | Literature & Fiction | World Literature             |
| 111-1-1111-1120-0 | Perfect Plants                          | B.M.Harwani | Home & Garden        | Crafts & Hobbies             |
| 111-1-1111-1121-0 | How to Grow in Small Space              | B.M.Harwani | Home & Garden        | Crafts & Hobbies             |
| 111-1-1111-1122-0 | Lifetime Gardening                      | B.M.Harwani | Home & Garden        | Antiques & Collectibles      |
| 111-1-1111-1123-0 | Home keeping A Must Have                 | B.M.Harwani | Home & Garden        | Antiques & Collectibles      |
| 111-1-1111-1124-0 | Designing A Home                        | B.M.Harwani | Home & Garden        | Interior Design              |
| 111-1-1111-1125-0 | Gardening A Complete Guide              | B.M.Harwani | Home & Garden        | Interior Design              |
| 111-1-1111-1126-0 | Growing Herbs At Home                   | B.M.Harwani | Home & Garden        | Home Design                  |
| 111-1-1111-1127-0 | Master Craftsman                        | B.M.Harwani | Home & Garden        | Home Design                  |
| 111-1-1111-1131-0 | Computer & Common Sense                 | B.M.Harwani | Computers & Internet | General                      |
| 111-1-1111-1131-1 | Introduction to Computers               | B.M.Harwani | Computers & Internet | General                      |
| 111-1-1111-1131-2 | Computer for Beginnners                 | B.M.Harwani | Computers & Internet | General                      |
| 111-1-1111-1131-3 | Visual Basic Programming                | B.M.Harwani | Computers & Internet | Programming                  |
| 111-1-1111-1131-4 | C# Programming                          | B.M.Harwani | Computers & Internet | Programming                  |
| 111-1-1111-1131-5 | C++ for Beginners                       | B.M.Harwani | Computers & Internet | Programming                  |
| 111-1-1111-1131-6 | Programming in C                        | B.M.Harwani | Computers & Internet | Programming                  |
| 111-1-1111-1131-7 | Mastering SQL Server                    | B.M.Harwani | Computers & Internet | Databases                    |
| 111-1-1111-1131-8 | Database Handling in Oracle             | B.M.Harwani | Computers & Internet | Databases                    |
| 111-1-1111-1131-9 | Practical Web Services                  | B.M.Harwani | Computers & Internet | Web Development              |
| 111-1-1111-1132-0 | jQuery Recipes                          | B.M.Harwani | Computers & Internet | Web Development              |
| 111-1-1111-1132-1 | Practical ASP.NET 3.5 Projects          | B.M.Harwani | Computers & Internet | Web Development              |
| 111-1-1111-1132-2 | Practical EJB Project                   | B.M.Harwani | Computers & Internet | Web Development              |
| 111-1-1111-1132-3 | Foundation Joomla                       | B.M.Harwani | Computers & Internet | Web Development              |
| 111-1-1111-1132-4 | Web Development with AJAX               | B.M.Harwani | Computers & Internet | Web Development              |
| 111-1-1111-1132-5 | Learn Mac OS X                          | B.M.Harwani | Computers & Internet | Operating Systems            |
| 111-1-1111-1132-6 | Learn Adobe Practically                 | B.M.Harwani | Computers & Internet | Graphic Design               |
| 111-1-1111-1132-7 | Secure your Hardware & Network          | B.M.Harwani | Computers & Internet | Security & Encryption        |
| 111-1-1111-1132-8 | Software Designing                      | B.M.Harwani | Computers & Internet | Software                     |
| 111-1-1111-1132-9 | Assemble your PC                        | B.M.Harwani | Computers & Internet | Hardware                     |
| 111-1-1111-1133-0 | iPhone SDK Programming Quickly & Easily | B.M.Harwani | Computers & Internet | Mobile & Wireless Computing  |
+-------------------+-----------------------------------------+-------------+----------------------+------------------------------+
```

Figure 1.5. Sample records in *books* table

We have examined the structure of the database tables that will be used in our web application. Let's now take a look at the actual screens the end user will see when our application is run.

1.4 Sample Output Screens

Our Book Store application will:

- Create a customer account
- Display books within categories and subcategories
- Add, delete, or update quantities of selected books in the cart
- Authenticate the user before actually placing the order
- Confirm the order with a Thank You message

The first screen the user will see when our program runs is the *Home* panel shown in Figure 1.6(a). The *Home* panel displays a Cart icon at the top with *0 items* and *0$*, indicating that the cart is empty. The *Home* panel also lists the menu items *Books, Contact Us, New Arrivals*, Discount Offers, Best Selling, and Gift Cards. Tapping a list item will display relevant information. Because we are trying to sell books, clearly the most important list item in the *Home* panel is the first one, *Books*.

1.4.1 Viewing Categories and Subcategories

When a user taps the *Books* list item in the *Home* panel, a list of categories from the *books* table is displayed, as shown in Figure 1.6(b). From the *Categories* panel, the user can return to the *Home* panel by tapping the *Home* button in the toolbar. If the user selects a category, the *Subcategories* panel appears. The list of subcategories that will appear if the *Literature & Fiction* category is selected is shown in Figure 1.6(c).

Figure 1.6. (a) *Home* panel screen (b) *Books* categories (c) List of subcategories

From the *Subcategories* panel, the user can return to the *Categories* panel by selecting the *Back* button from the toolbar. The user can also select a subcategory to see a list of books. The list of books shown in the *Select Books* panel is fetched from the *books* table of our *shopping* database. Multiple books display as a scrollable list, as shown in Figure 1.7 (b).

1.4.2 Viewing Detailed Information and Adding Items to the Cart

Every book displayed in the *Select Books* panel is has a *Quantity* input field and two buttons: *Show Details* and *Add To Cart*. Assuming that the customer will buy at least one copy of the book, the default value of the *Quantity* field is set to *1,* as shown in Figure 1.7(a). The customer can, of course, change the quantity and tap the *Add To Cart* button to drop them into the Shopping Cart.

9

Tapping the *Show Details* button will display details and a short description of the content, as shown in Figure 1.7(c).

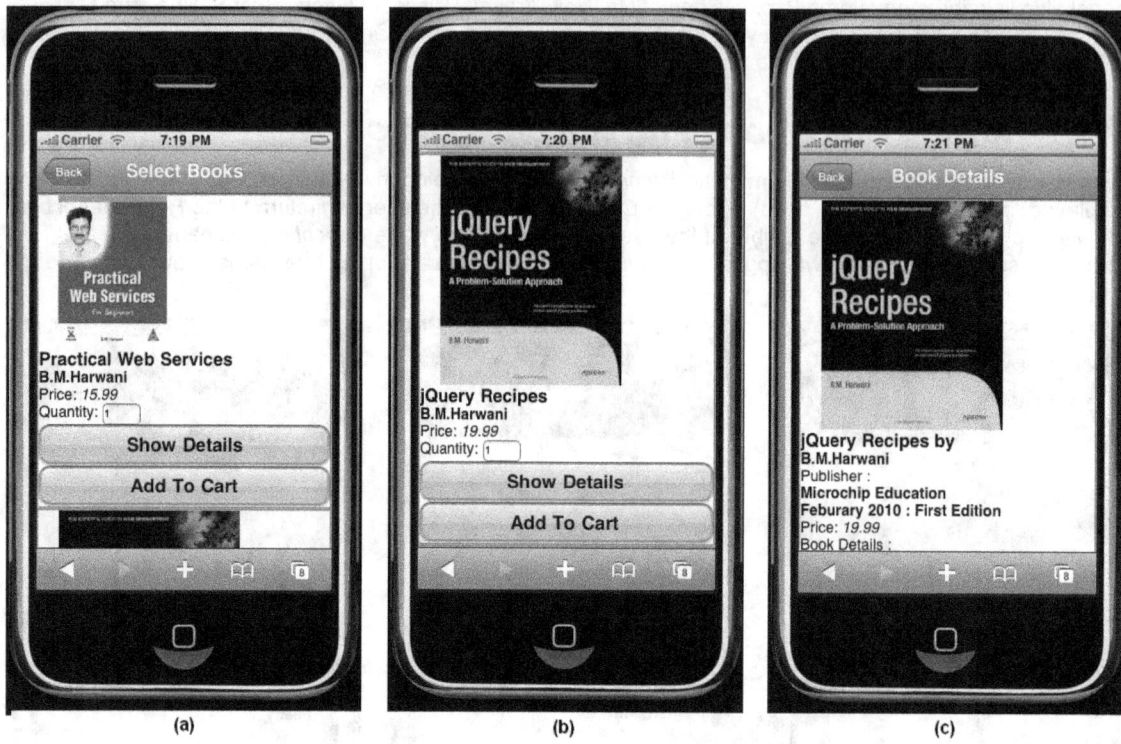

(a)   (b)   (c)

Figure 1.7. (a) List of books displayed in a subcategory (b) Scrolling the screen to view more books

(c) Detailed information via the *Show Details* button

The user may need to scroll the screen to see hidden information. Figure 1.8(a) shows a brief description, an input field to specify the number of copies, and an Add to Cart button. Let's pretend that our user is extraordinarily discerning and wishes to buy three copies of the selected book. Figure 1.8(b) shows the input field. When the *Add to Cart* button is tapped, the book is inserted in the cart and we will be jump to the *Cart Updated* panel, as shown in Figure 1.8(c).

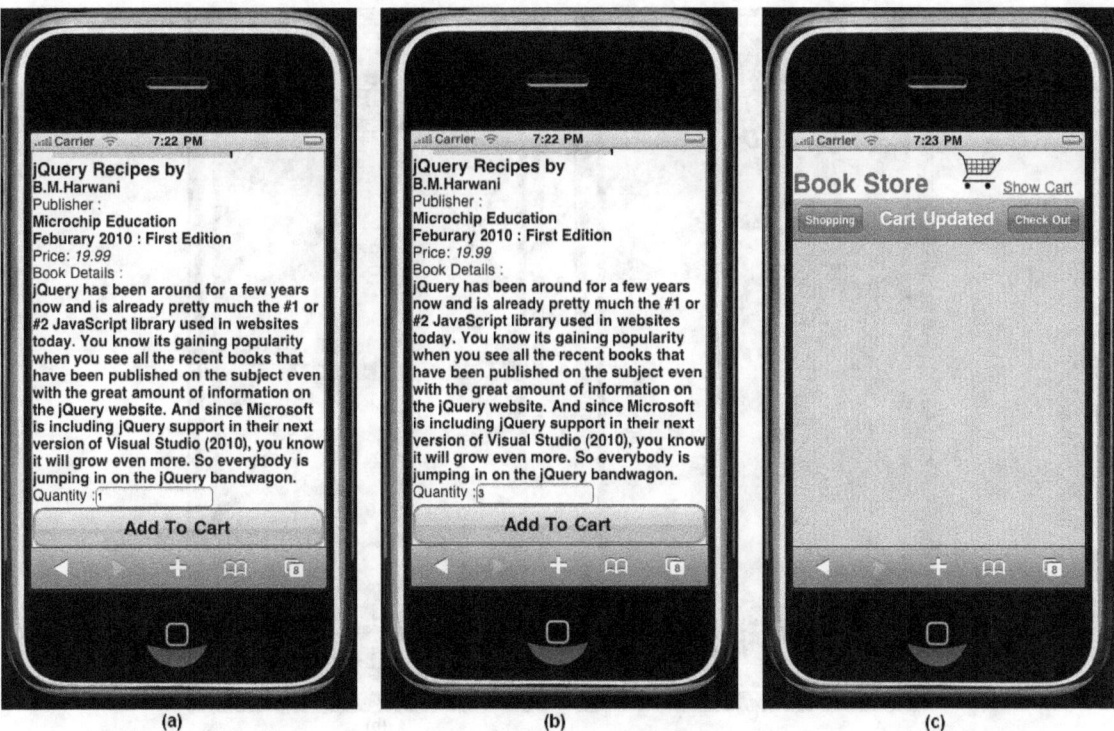

(a) (b) (c)

Figure 1.8. (a) Detailed information appears when the screen is scrolled (b) Specifying the book quantity

(c) Screen confirming that the Cart has been updated

The *Cart Updated* screen also displays a *Show Cart* link next to the cart icon, and two toolbar buttons: *Shopping* and *Check Out,* allowing the customer to continue shopping or pay for the merchandise.

1.4.3 Maintaining the Cart

If the user selects the *Show Cart* link, the *Items in Cart* panel appears, displaying the list of books present in the cart. In our sample, we'll see 3 copies of *jQuery Recipes*, as shown in Figure 1.9(a). We want to give the customer a chance to change the quantity (or delete the book altogether), so this panel displays two buttons: *Delete* and *Update*, for every item in the cart.. Let's reduce the quantity of the book from *3* to *1,* as shown in Figure 1.9(b). After changing the value in the *Quantity* input field, the user can select the *Update* button to recalculate the cart. After selecting the *Update* button, the application jumps to the *Cart Updated* panel, as shown in Figure 1.8(c).

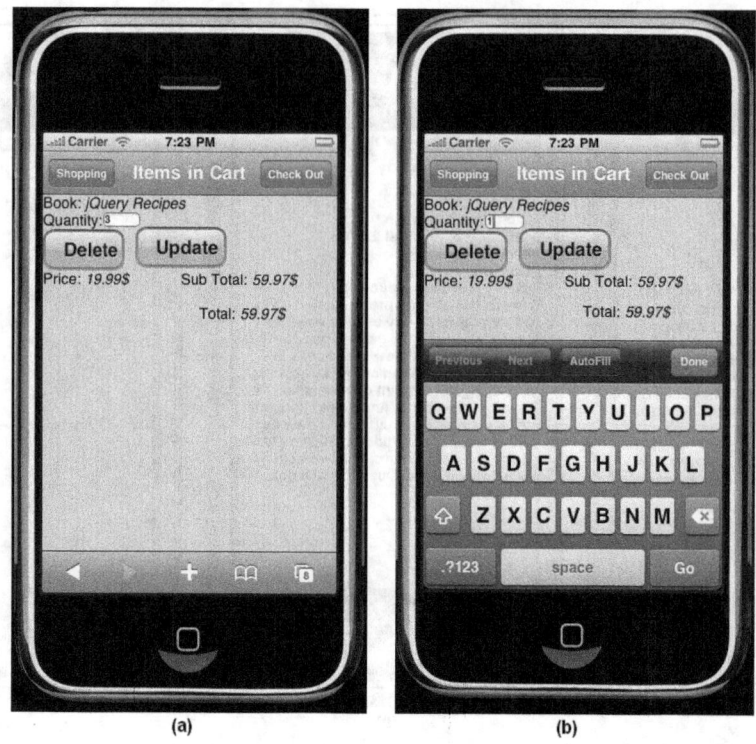

(a)　　　　　　　　　　(b)

Figure 1.9. (a) Items inserted in the Cart displayed when the Show Cart link is selected

(b) Updating quantity

Figure 1.10 (a) shows the new quantity of *1*. This time, let's select the *Shopping* button from the toolbar to buy more of my well-written, insightful, and educational books. When the *Shopping* button is tapped, the *Categories* panel is redisplayed, as shown in Figure 1.10(b). Notice that the shopping cart link is no longer empty—it contains one item and a total price of $19.95. The user can select this link and peruse the complete cart contents, as shown in Figure 1.10(a), or return to *Home* panel by selecting the *Home* button from the toolbar, as shown in Figure 1.10(c). Notice that the shopping cart link is displayed in this panel, as well.

Figure 1.10. (a) Cart with updated quantity (b) *Category* panel with filled shopping cart (c) *Home* panel with filled shopping cart

Let's do a little more shopping by selecting the *Home & Garden* category from the *Categories* panel (see Figure 1.11(a), then display books in the *Crafts & Hobbies* subcategory, as shown in Figure 1.11(c).

Figure 1.11. (a) Selecting the *Home & Garden* category (b) Selecting the *Crafts & Hobbies* subcategory (c) Viewing the list of books in the *Crafts & Hobbies* subcategory

Leaving the default quantity as 1, let's add the book to the cart by selecting the *Add to Cart* button. The book is added to the cart and we will jump to the *Cart Updated* panel. To see the current cart contents, select the *Show Cart* link and we'll jump to the *Items in Cart* panel (see Figure 1.12). The cart now contains two books with one copy each.

Let's delete *jQuery Recipes* from the cart by selecting the *Delete* button. When this button is tapped, the book is removed from the cart the *Cart Updated* panel appears with modified cart, which now contains a single book title, *Perfect Plants*.

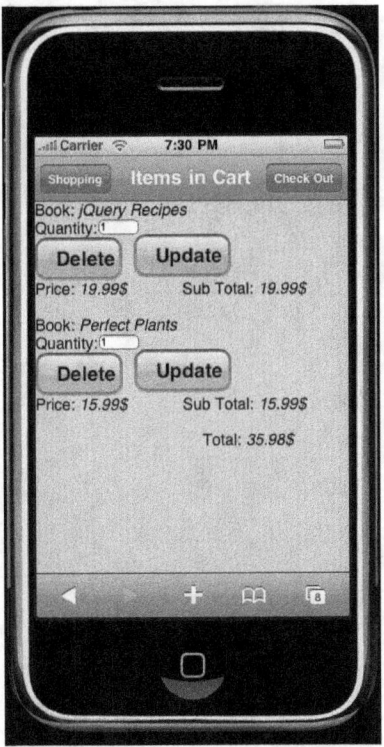

Figure 1.12. List of books in the Cart

To actually place the order, the user must select the *Check Out* button from the toolbar. But before the merchandise can be paid for, the user has to create and account and enter some information.

1.4.4 Creating an Account and Signing In

When the *Check Out* button is tapped, the *Checking Out* panel appears, prompting previous users to sign in or new users to create an account, as shown in Figure 1.13(b). When the *Sign In* button is chosen, the *Sign In* panel appears, asking for a *userid* and *password,* as shown in Figure 1.13(c). The two input fields contain light gray placeholder text by default, which guides the user in filling out their information.

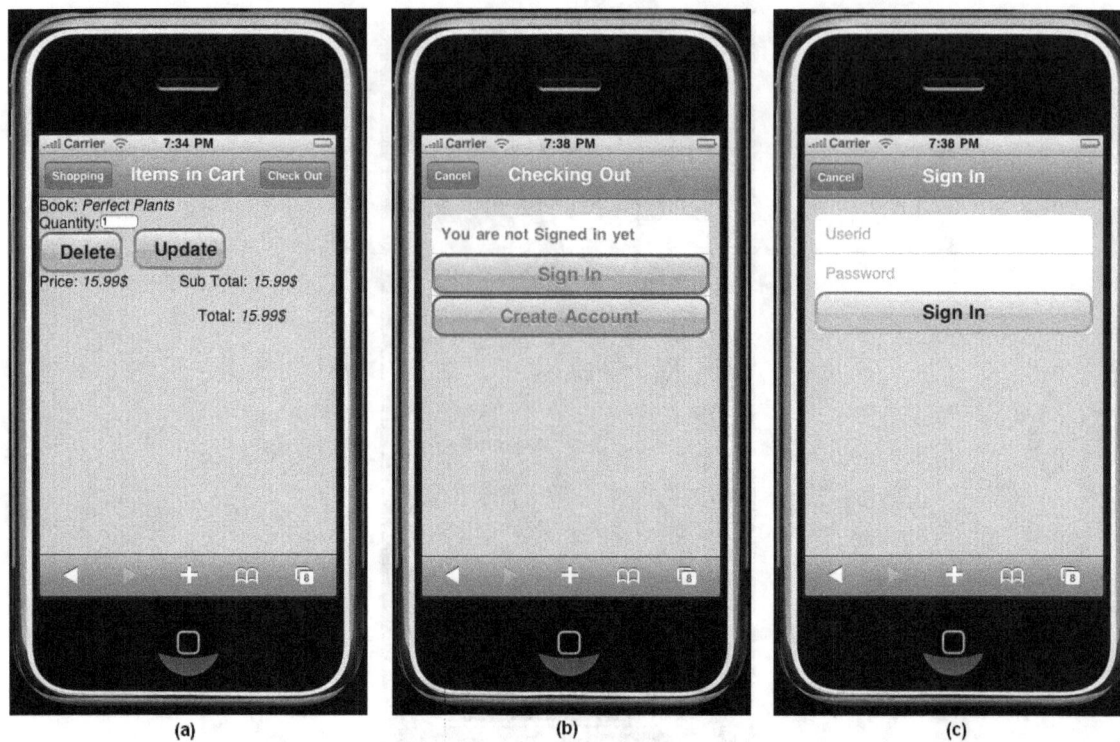

(a)   (b)   (c)

Figure 1.13. (a) Book(s) present in the Cart after the Show Cart link is selected (b) Checking Out screen (c) Sign In screen

Validation checks are applied to the *userid* and *password* fields; neither field may be empty. If the user selects the *Sign In* button without entering a user id, an error message appears, as shown Figure 1.14(a). If the password field is blank, a similar error message is displayed, as shown in Figure 1.14(b). If the user supplies incorrect data, the message *Sorry the userid or password is incorrect* appears*, with two buttons, Try Again* and *Create Account*, allowing the user to re-enter either the user id and/or password, or create a new account, as shown in Figure 1.14(c).

Figure 1.14. (a) Blank User id error message (b) Blank Password message (c)Incorrect User id or Password message

If the user selects the Cre*ate Account* button, the page switches to the *Create Account* panel, as shown in Figure 1.15 (a). Each input text field contains light gray placeholder text by default. There are quite a few fields in this panel, so the user will have to scroll down to see all of them, as shown in Figure 1.15(b). The last field (*Country*) is followed by a *Submit* button, which the user will tap after filling in information in the respective fields.

Validation checks are applied to all these fields and none of the essential ones, such as *userid, password, name, address, email address* or *contact number* may be left blank; otherwise an error message will be displayed, as shown in Figure 1.15(c).

(a)   (b)   (c)

Figure 1.15. (a) Top half of the *Create Account* screen (b) Bottom half of the *Create Account* screen (c) Error message produced if *User id* is left blank

If the passwords entered in the two fields, *password* and *Re-enter password,* are different, another error message is displayed: *Password and Re-enter password don't match. Please enter again,* as shown in Figure 1.16(a). Finally, if all the data is entered correctly and the Submit button pressed, a user account is created. Underneath the resulting Congratulations message, a *Sign In* button is presented, allowing the user to actually sign in, as shown in Figure 1.16(c).

Note: The information supplied by the user will be stored in the *customers* table of the *shopping* database.

18

(a)	(b)	(c)

Figure 1.16. (a) Error message displayed when the passwords don't match (b) Correctly entered user information (c) Congratulations message after successful creation

When the user taps the *Sign In* button, the *Sign In* panel appears (see Figure 1.17(a), and, after entering the correct user is and password, a Welcome message appears (see Figure 1.17(b). A *Supply Shipping Info* button is also displayed so the user can add shipping information.

1.4.5 Supplying Shipping Information

After selecting the *Supply Shipping Info* button, the *Placing Order* panel is displayed. The text fields have already been filled in with the information supplied during account creation and stored in the *customers* table, as shown in Figure 1.17(c). This information is not editable. Instead, a second set of fields appears below the original data, where the user may enter a shipping address and credit card information (see Figure 1.18(a).

Finally, after much tapping and form-filling, the user is prompted to tap the *Place Order* button, after which a message is displayed that thanks the user and provides the order number, as shown in Figure 1.18(c). Note that the cart information will be stored in both the *orders* and *orders_details* tables.

19

Figure 1.17, (a) Sign In screen (b) Welcome message (c) Shipping Information

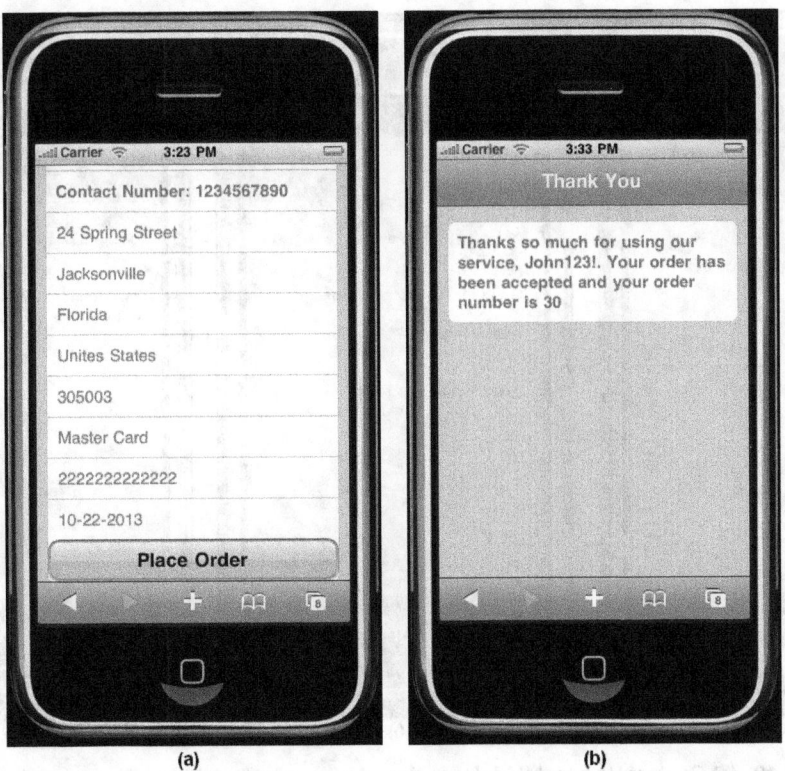

(a)   (b)

Figure 1.18. (a) Fields with completed shipping information (b) Thank You and order number message

1.5 A Few Final Details

The remainder of the list items in the *Home* panel (see Figure 1.6 (a) are minor, but necessary, displaying single screens of related information. The *Contact Us* list item shows the address and email of our dummy store (see Figure 1.19(a)). *New Arrivals* displays newly arrived books (see Figure 1.19 (b)). And the *Deep Discount* screen shows a list of discounted titles (see Figure 1.19(c)).

Figure 1.19   (a) Contact Us panel (b) Information of new arrivals (c)Deep Discount information

1.6 Summary

In this chapter, we briefly discussed PHP, MySQL, and the role of a web server in the execution of a web application. We explored the structure of the four tables used in our application. We also examined the sample output of our web application. In the next chapter, we will introduce jQTouch and begin developing our application.

2

Starting with jQTouch

In the previous chapter, we inspected the outline of the web application we're going to build in this book. We also saw the structure of different database tables we'll be using in this application. In this chapter, we're going to:

- Learn how a small application can be built using jQTouch

- Learn how to change themes, add panels, display titles automatically, and highlight certain information

- Learn how to get user information through a form

- Learn how to display user information and apply validity checks to the input fields

2.1 Introduction to jQTouch

jQTouch is a jQuery plugin for mobile web development created by David Kaneda that can be used for building iPhone, Android, and iPad applications. jQTouch is targeted to small screen devices and makes programming for mobile browsers easier, providing native animations, auto-list navigation, and default application styles for WebKit-based mobile browsers like the iPhone/iPod Touch, Android, and Palm webOS.

Without jQTouch, we would expend a great deal of useless effort creating HTML styles so that applications would mimic the mobile screen. Not only this, we would have to write code to:

- Make the web pages resemble iPhone screens, with a gradient background, toolbar, and so on.

- Add a standard iPhone-esque *back* button to the application, which, when selected, returns to the previous screen. And we would therefore have to write code that logs the user's click history.

- Create different animation effects while navigating from one screen to another, such as iPhone-like slide effects that mimic a left-sliding page when selecting an item and a right-sliding page when selecting *back* button.

- Truncate long header lines a with trailing ellipsis.

jQTouch does all this automatically, making app development faster and easier. So, to get started, download the latest version from *http://jqtouch.com/*. Unzip the jQTouch in the same directory where the web app is being developed. We will be testing our web applications through Apache web server and the folder where Apache Web Server looks for web app files is */Library/WebServer/Documents* folder. So, lets unzip jQTouch also in the same folder. Let's get started

Note: I asssume, you have PHP, MySQL and Apache Web Server installed and configured on your machine. Please refer Appendix A for any help

2.2 Creating Your First Application

The approach for developing applications in jQTouch is very simple. The steps are:

- For every screen of our web app, create a separate div element and assign it a unique id. A *div* element is a popular HTML element heavily used for laying out web pages. A div also helps in defining a division or section of an html document.

- Create an unordered list in the home screen div element. Then, in each list item, place an anchor tag with an href that links to other divs (screens) of the application.

jQTouch will hide all the divs except the home screen, and render our Home Screen list items with attractive gradients. Tapping a list item causes the home screen to slide off and the linked div element to slide on. Hence, by using jQTouch, we can easily create and format the various screens of our web app and apply animated transitions between them.

Note: jQTouch takes any direct descendent in the <body> section and converts it into a full-screen panel that can be animated in either direction.

We will make three screens for our Book Store: a *Home* page, a *Books* page, and a *Contact Us* page. The three pages are represented by three div elements that are assigned the unique ids *home, books* and *contactus* respectively. Each div element that is a direct descendant of the body section will become a panel in the application. The two panels, *books* and *contactus,* additionally contain paragraph elements for displaying information on their respective pages.

Our three screens are shown in Figures 2.1(a), 2.1(b), and 2.1 (c). And we will be calling these three main div elements as *panels* from now on. Each panel will also contain a toolbar for displaying the panel title and navigation buttons. The complete code of the application is shown in Listing 2.1.

Note: To try the examples of this book create them in */Library/WebServer/Documents* folder. If the following application is created and saved by name say a1.html, then to test it, open the browser and point it at the address *http://localhost/a1.html*

Listing 2.1. Initial Application Code

```
<html>

    <head>

        <title>Books Store</title>

        <link type="text/css" rel="stylesheet" media="screen" href="jqtouch/jqtouch.css">

        <link type="text/css" rel="stylesheet" media="screen" href="themes/jqt/theme.css">

        <script type="text/javascript" src="jqtouch/jquery.1.3.2.min.js"></script>

        <script type="text/javascript" src="jqtouch/jqtouch.js"></script>

        <script type="text/javascript">

            var jQT = new $.jQTouch();

        </script>

    </head>

    <body>

        <div id="home">

            <div class="toolbar">

                <h1>Book Store</h1>

            </div>

            <ul class="edgetoedge">

                <li class="arrow"><a href="#books">Books</a></li>

                <li class="arrow"><a href="#contactus">Contact Us</a></li>

            </ul>

        </div>
```

```
    <div id="books">

        <div class="toolbar">

            <h1>Books</h1>

            <a class="button back" href="#">Back</a>

        </div>

        <p>We are a US-based organization providing a wide variety of books at a

            reasonable price</p>

    </div>

    <div id="contactus">

        <div class="toolbar">

            <h1>Contact Us</h1>

            <a class="button back" href="#">Back</a>

        </div>

        <p>XYZ Book Company</p>

        <p>11 Books Street, NY, NY 10012 </p>

        <p>USA</p>

    </div>

  </body>

</html>
```

The first thing we do in Listing 2.1 is link the CSS and Javascript files by adding references to the jQTouch stylesheet and themes. The style rules in the *jqtouch.css* file are used for handling animations, orientations, and so on. Also, we add references to the appropriate Javascript files—*jquery.1.3.2.min.js* and *jqtouch.js*. Remember that jQTouch requires, and is bundled with, its own copy of jQuery (though we can link another copy if desired). For the application to work, the reference to jQuery must be added before the reference to jQTouch.

As far as themes are concerned, jQTouch requires one basic theme to make page transitions work and provide iPhone native screens. jQTouch provides two themes:

- apple, which mimics the default iPhone look
- jqt, which is based on apple but is darker and more universal

We can also create our own custom themes either by building them from scratch or by modifying existing jQTouch themes.

After linking the CSS and Javascript files, we need to initialize the jQTouch object, as shown here:

```
<script>

    var jQT = new $.jQTouch();

</script>
```

This function initializes a jQTouch object and assigns it to variable *jQT*. The variable *jQT* is then used to manipulate jQTouch with JavaScript through its functions. We can also specify certain optional property values while initializing the jQTouch object, enabling us to customize the behavior of our web application. We will examine these properties later in this chapter.

As we said earlier, all the screens of a web application are represented by individual div elements. All panels have toolbars, which are divs of class *toolbar*. The toolbar is displayed at the top of the every panel and appears as in the traditional iPhone format. A toolbar always contains an *h1* element that represents the *title* of the panel and two optional buttons, one of which can be a *back* button. The *back* button appears on the left side of the toolbar and is used for navigating to the previous page.

We have added buttons with *button* and *back* class names in the toolbar of both the *Books* and *Contact Us* *pages*. The *href* on the *back* button is set to *#,* which will navigate to the previous panel. We can, of course, navigate to a panel with a specific id instead, by setting the *href* to point at it. For example, if the *href* of the *back* button is set to *#home,* we will always be sent to *home.* The two list items *(li)* in the home panel have hyperlinks pointing to *books* and *contactus,* which means that when these list items are selected, we will jump to the panel of the respective id.

We want the *home* page to display navigation links to the *Book* and *Contact Us* pages. To display such navigation links, we use an unordered list. We can add simple classes to the *<ul>* elements to change their style, for example, *edgetoedge* or *rounded:*

- The *edgetoedge* class makes the unordered list appear in the iPhone's edge-to-edge navigation list design. Each of the items in *<ul>* element of class *edgetoedge* is stretched from left to right in the viewable area for easy touch input, as shown inFigure 2.2(a).

- The *rounded* class is usually used for destination pages to show an information list. The items in a *rounded* class *<ul>* element are surrounded by a rounded rectangle box, as shown in Figure 2.3(a).

Each item within an unordered list contains a hyperlink that navigates to the designated application panel. We may optionally apply the *arrow* class to the list items, which will display an arrow to the right of the item. The icon tells the user that there is an additional web page (screen) to be viewed.

Whenever we navigate to a new page, two animations occur: The new page *animates in* and the old page *animates out.* When we select a *back* button, *reverse animation* takes place. When a list item is selected, the default animation makes it appear as if the new page is sliding in from the right side of the screen. We can move to another web page by applying any of the eight different animation effects: slide, slideup, dissolve, fade, flip, pop, swap, and cube.

Note: The toolbar button may be of any class—*back*, *button* or *cancel*—but may have any animation style.

Each visited screen is inserted into the *page history*, and we can navigate back to any previous page of the history by using the *back* button.

When the code in Listing 2.1 is executed, the first panel, *home,* appears, as shown in Figure 2.1(a). The panel title is *Book Store*, which is followed by two list items, *Books* and *Contact Us*. When the *Books* list item is clicked, we jump to the *books* panel, as shown in Figure 2.1(b). The title of the target panel is *Books* and there's a *Back* button on the toolbar. The *books* panel information appears below the toolbar. If we select the *Back* button, we will return to the *home* panel.

Similarly, when *Contact Us* is selected, we jump to the *contactus* panel, as shown in Figure 2.1(c).

Note: Since we have not yet specified the type of animation in the anchor tags, the default animation, *slide,* is applied.

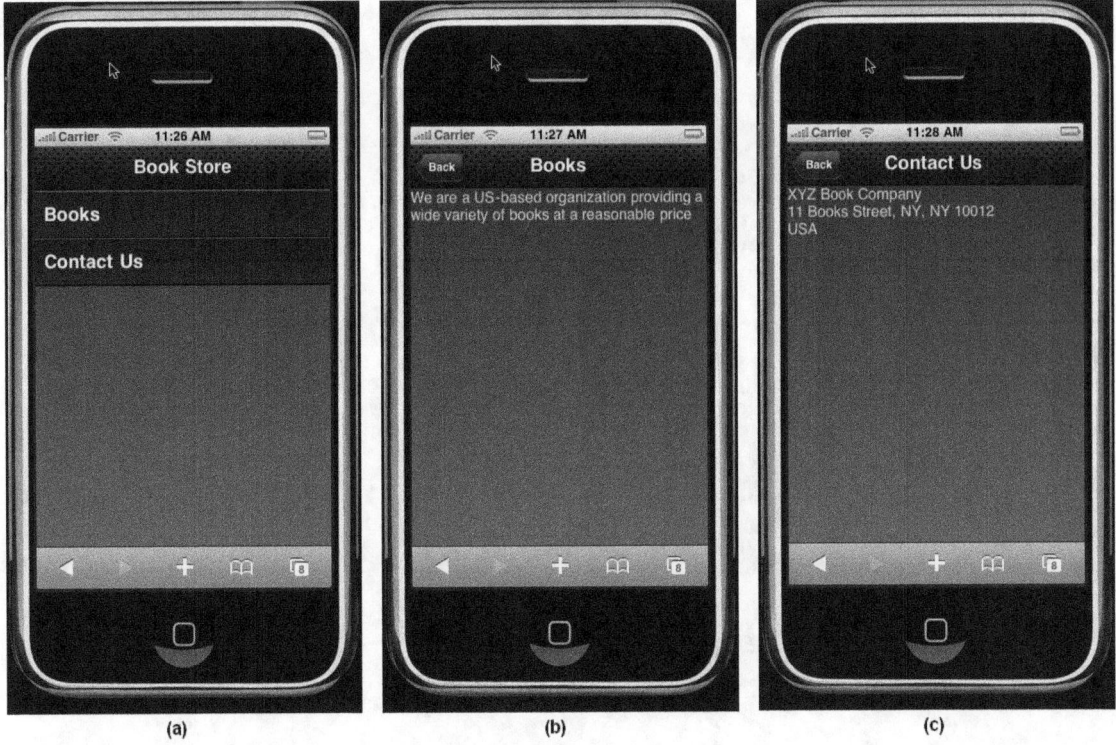

(a)   (b)   (c)

Figure 2.1. (a) *Home* panel with two list items (b) *Books* panel with *back* button (c) *Contact Us* panel with *back* button

2.3 Applying The Apple Theme

Our first round of coding used the *jqt* theme, which is fairly dark. Let's see how our page looks when the theme is changed to *apple*. This is simple to do—we just need to replace the word *jqt* with *apple* in the line that adds reference to the jQTouch theme. Thus:

```
<link type="text/css" rel="stylesheet" media="screen" href="themes/jqt/theme.css">
```

is replaced by:

```
<link type="text/css" rel="stylesheet" media="screen" href="themes/apple/theme.css">
```

Now our application will appear in the *apple* theme, as shown in Figures 2.2(a), 2.2(b), and 2.2(c).

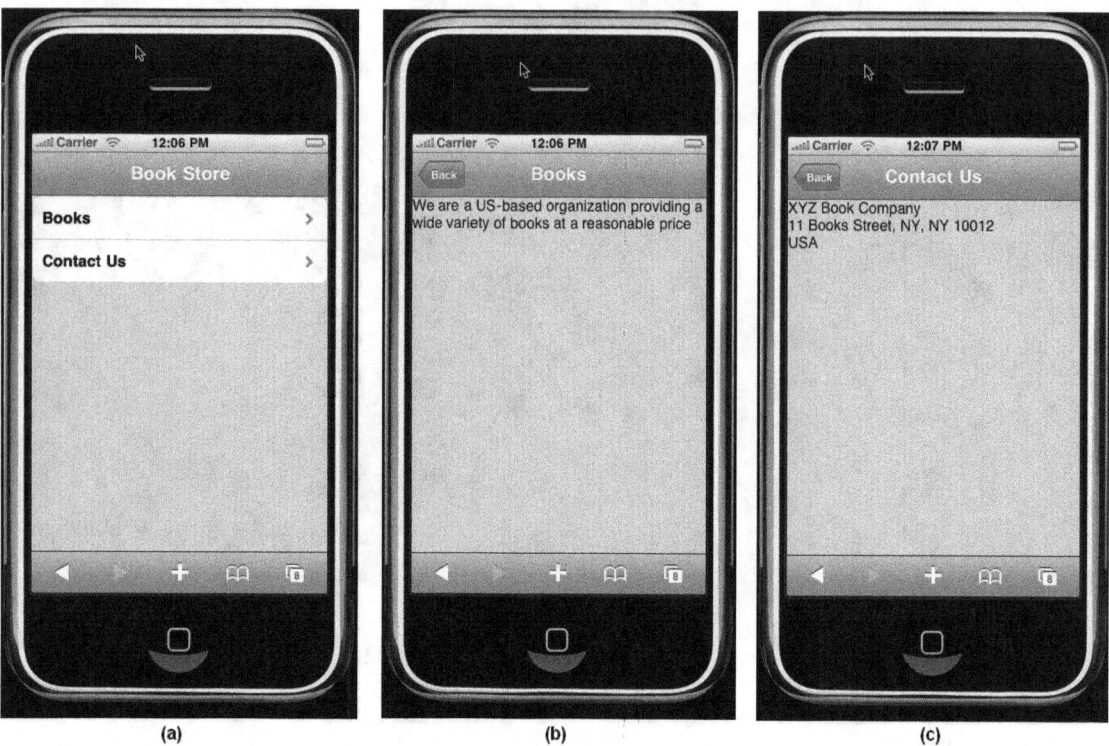

Figure 2.2.(a) *Home* **panel with two list items (b)** *Books* **panel with** *back* **button (c)** *Contact Us* **panel with** *back* **button**

So far so good. But our application is a bit underpowered. It doesn't shows different book categories. Also, there's no email link provided in the Contact Us panel. So, let us go ahead and beef up our application a bit.

2.4 Adding More Panels to the Application

In this section, we will add an additional panel to our application that displays book categories when the *Books* list item is selected from the *home* panel. Also, we will add an *email* link in *Contact Us* panel. For the time being, we'll hard-code the book categories, but in later chapters, we'll learn how to fetch the categories from database tables placed on the server.

Let's start by modifying our application to display different categories via an unordered list. Wrapped inside unordered list will be the list items, each representing a category. To provide navigation, each list item will contain a hyperlink pointing to a new panel, *book,* that will display the information of the selected book category,

The code for the modified application is shown in Listing 2.2. The bold code shows the changes.

Listing 2.2. Adding Another Panel to the Application

```html
<html>

    <head>

        <title>Books Store</title>

        <link type="text/css" rel="stylesheet" media="screen" href="jqtouch/jqtouch.css">

        <link type="text/css" rel="stylesheet" media="screen"
```

```
        href="themes/apple/theme.css">
    <script type="text/javascript" src="jqtouch/jquery.1.3.2.min.js"></script>
    <script type="text/javascript" src="jqtouch/jqtouch.js"></script>
    <script type="text/javascript">
        var jQT = new $.jQTouch();
    </script>
</head>
<body>
    <div id="home" >
        <div class="toolbar">
            <h1>Book Store</h1>
        </div>
        <ul class="rounded">
            <li class="arrow"><a href="#books">Books</a></li>
            <li class="arrow"><a href="#contactus">Contact Us</a></li>
        </ul>
    </div>
    <div id="books">
        <div class="toolbar">
            <h1>Categories</h1>
            <a class="button back" href="#">Back</a>
        </div>
        <ul class="rounded">
            <li class="arrow"><a href="#book">Literature & Fiction</a></li>
            <li class="arrow"><a href="#book">Home & Garden</a></li>
            <li class="arrow"><a href="#book">Computers & Internet</a></li>
            <li class="arrow"><a href="#book">Cooking, Food & Wine</a></li>
        </ul>
    </div>
    <div id="book">
        <div class="toolbar">
            <h1>Books</h1>
            <a class="button back" href="#">Back</a>
        </div>
        <p>Following is the list of books available in this category</p>
    </div>
    <div id="contactus">
        <div class="toolbar">
```

```
            <h1>Contact Us</h1>

            <a class="button back" href="#">Back</a>

        </div>

        <p>XYZ Book Company</p>

        <p>11 Books Street, NY, NY 10012 </p>

        <p>USA</p>

        Email us: <a href="mailto:bmharwani@yahoo.com"

            target="_blank">bmharwani@yahoo.com</a>

    </div>

  </body>

</html>
```

Note: This time, we applied the *rounded* class to the unordered list, which will give us a chance to see how the appearance of the application changes compared to the *edgetoedge* class used in Listing 2.1.

In Listing 2.2, The *books* panel was created via a *books* div element, then modified to display an unordered list. The *rounded* class was applied to the unordered list, and four book categories were added. Each list item category contains a hyperlink pointing to the newly added b*ook* panel.

For now, the *book* panel only contains a paragraph element displaying the text message: *Following is the list of books available in this category.* In later chapters, we'll learn how to fetch books in a selected category from server-side database tables and display them in this panel.

In the *contactus* panel, we added a line *Email us* below the paragraph elements. The code contains a hyperlink which, when selected, opens a screen that allows the user to enter a subject and text and send it to the specified email address.

Let's run the application to see how our improved application appears. The first screen is the *home* panel shown in Figure 2.3(a). When the *Books* list item is selected, we jump to the *Categories* panel, which we have modified to display book categories, as shown in Figure 2.3(b). When a category is selected, we'll jump to the newly added *Books* panel, which displays the message, *Following is the list of books available in this category*, as shown in Figure 2.3(c). No books are displayed below this message at the moment, but in later chapters, we'll learn to fetch and display the related books from the server-side database tables.

Figure 2.3. (a) The *Home* panel with two list items (b) The *Categories* panel with four list items

(c) Message displayed in the *Books* panel

When the Back button is selected from the *Books* panel toolbar, we will be returned to the *Categories* panel. Tapping the Back button here returns us to the *home* panel.

When we select the *Contact Us* list item in the *home* panel, the *contactus* panel appears, displaying the contact information of our dummy organization, as shown in Figure 2.4(a). Notice we now have an *email link*. Tapping this link brings up a *New Message* page, as shown in Figure 2.4(b).

Figure 2.4. (a) Information displayed in the *Contact Us* panel (b) *New Message* panel

At the moment, we don't have any category-specific pages—tapping any category brings up the same plain-vanilla page. So our next task is to add pages that reflect the currently chosen category.

2.5 Automatically Displaying Titles

To automatically display the appropriate panel title, we'll make use of *extensions* provided by jQTouch. To do this, we need to add a reference to the javascript file, *autotitles.js*. Let's add the following line to the head element of our application:

```
<script src="extensions/jqt.autotitles.js" type="application/x-javascript" charset="utf-8"></script>
```

The head element of our application now looks like this:

```
<head>

    <title>Book Store</title>

    <link type="text/css" rel="stylesheet" media="screen" href="jqtouch/jqtouch.css">

    <link type="text/css" rel="stylesheet" media="screen" href="themes/apple/theme.css">

    <script type="text/javascript" src="jqtouch/jquery.1.3.2.min.js"></script>

    <script type="text/javascript" src="jqtouch/jqtouch.js"></script>

    <script src="extensions/jqt.autotitles.js" type="application/x-javascript"
        charset="utf-8"></script>

    <script type="text/javascript">
```

32

```
        var jQT = new $.jQTouch();

    </script>

</head>
```

Let's run the application to see what happens. When the *Literature & Fiction* category is selected from from the *Books* panel, we jump to the *book* panel. The title now reads *Literature & Fiction*, as shown in Figure 2.5(a). Similarly, when the category *Home & Garden* is chosen, the title of the *book* panel changes to *Home & Garden* as shown in Figure 2.5(b).

Note: If a heading is too long to fit in the allotted space, ellipses appear at the end of the text.

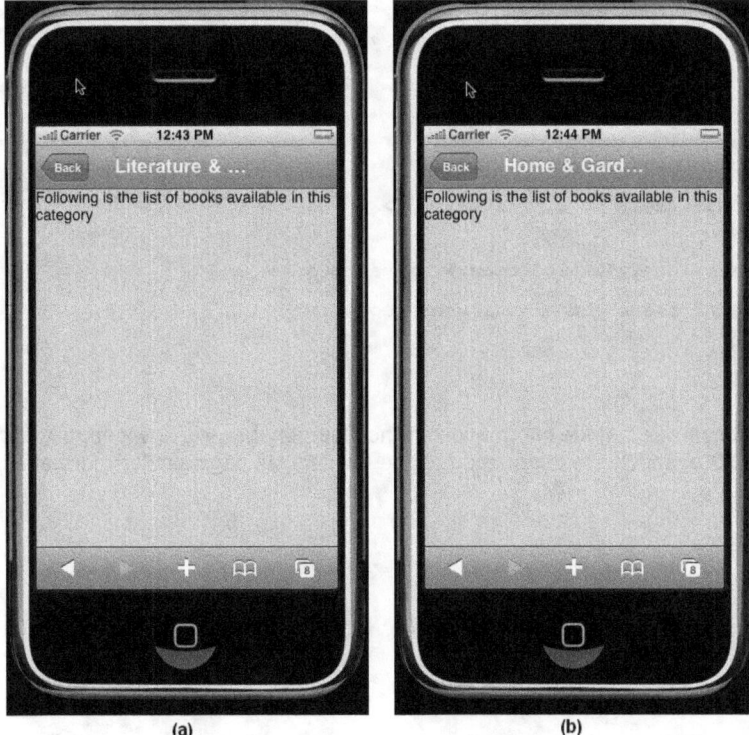

(a)   (b)

Figure 2.5. (a) The Category name when the Literature & Fiction category is selected. (b) The Category name when the Home & Garden category is selected.

2.6 Highlighting Information

Let's tweak the screen appearance a bit to centrally align and highlight specific information. jQTouch makes this task simple. All we have to do is wrap the data we wish to align and highlight inside the *div* element of class *info*. In this example, we'll wrap the paragraph elements of the *book* and *contactus* panels, as shown in the following code fragment. The new code is shown in bold; the remainder of the code is identical to Listing 2.2.

```
<div id="book">

    <div class="toolbar">

        <h1>Books</h1>

        <a class="button back" href="#">Back</a>

    </div>
```

```
<div class="info">

    <p>Following is the list of books available in this category</p>

</div>

</div>


<div id="contactus">

    <div class="toolbar">

        <h1>Contact Us</h1>

        <a class="button back" href="#">Back</a>

    </div>

    <div class="info">

        <p>XYZ Book Company</p>

        <p>11 Books Street, NY, NY 10012 </p>
        <p>USA</p>

        Email us: <a href="mailto:bmharwani@yahoo.com"

            target="_blank">bmharwani@yahoo.com</a>

    </div>

</div>
```

In Figure 2.6(a), you can see that the information on the Category page is now centrally aligned and highlighted. Figure 2.6(b) shows the information on the Contact Us page similarly formatted.

Figure 2.6. (a). Book information with *info* class applied (b) Contact Us information with *info* class applied

We have successfully aligned and highlighted specific information of our application. One really important feature is still missing in our application: user account information.

Most web applications have the ability to create user accounts. The information supplied by the user is needed for future authentication and fetching contact information, for example, the delivery address. User data is usually stored on server-side databases. Recall that we created a *shopping* database in Chapter 1 with four tables in it *books, customers, orders* and *orders_details*. The *shopping* database is an example of a server-side database, as it will be maintained on the server where our web application is hosted. In the next few sections, we'll learn how to:

- Get information from the user
- Display the user-entered information on the screen
- Apply validity checks to confirm that none of the essential fields has been left blank

2.7 Obtaining User Information

Let's add a panel to our application that asks the user to enter information such as user id, contact number, and email address.

We need to do two things to create a form:

- Add an additional panel to our application containing a form with text input fields
- Add a list item with a hyperlink to navigate to the newly added panel

Let's add a panel containing a form to the code shown in Listing 2.2. The panel will be used for creating an account, so let's call it *createacct*. The code for the newly added panel is shown here:

```
<div id="createacct">
```

```
<div class="toolbar">

    <h1>create account</h1>

    <a class="button cancel" href="#">Cancel</a>

</div>

<form id="newuser">

    <ul class="rounded">

        <li><input type="text" id="Userid" placeholder="Userid" /></li>

        <li><input type="text" id="ContactNo" placeholder="Contact No" /></li>

        <li><input type="text" id="EmailAddress" placeholder="Email address" />

            </li>

    </ul>

    <a href="#" class="whiteButton">create account</a>

</form>

<div>
```

So far, we've been using buttons with *button* and *back* class names in the toolbar. In the *createacct* panel, we've used buttons with *button* and *cancel* class names. The *cancel* button behave just like the *back* button, except that it removes the current page from view using reverse animation. So, if we navigate to the *createacct* panel with a *slide up* animation, when the *cancel* button is selected, we will navigate to the previous view via the *slide down* animation.

Because we have not specified any particular animation effects for hyperlinks, the default effect, *slide in,* is automatically applied to all page jumps. Thus, when the *cancel* button is tapped, we return to the *home* panel with the *slide out* effect. Remember that, unlike *back* buttons, *cancel* buttons are not left arrow icons.

In the previous code fragment, we have added a form and assigned it *newuser* id. The form contains an unordered list with three list items. Each list item contains a text field asking for information from the user. The input text fields are embedded inside the list items so that jQTouch's *apple* theme (or *jqt* theme) can be used to style the form. We have used three attributes: *type, id,* and *placeholder* for the input text fields. The complete list of attributes usable with input text fields is shown in Table 2.1.

Table 2.1. Input Text Field Attributes

Attribute	Usage
type	Set to *text* to specify a single-line text entry field.
placeholder	A string displayed in an empty text field that shows the user what needs to be entered
name	Specifies the user name to apply to the data when the form is submitted
id	Specifies a unique identifier for the text field
autocapitalize	Turns off the default auto-capitalization feature of Mobile Safari.
autocorrect	Turns off the default spell check feature of Mobile Safari.
autocomplete	Turns off the autocomplete feature of Mobile Safari.

Three input text fields will now appear on the relevant page with placeholders that will be replaced by user input. At the end of the unordered list containing the input text fields, there's a hyperlink of class *whiteButton,* which displays the button shown in Figure 2.7(b). This button contains the text *Create Account*. We'll soon learn how to initiate certain tasks when this button is selected by the user.

We need to add an extra list item in the *home* panel which, when selected, will navigate us to the newly added *Create Account* panel. The list item will include a hyperlink pointing to the id of our *Create Account* panel: *createacct* . The new list item is shown in bold in the following code fragment:

```
<div id="home">

    <div class="toolbar">

        <h1>Book Store</h1>

    </div>

    <ul class="rounded">

        <li class="arrow"><a href="#books">Books</a></li>

        <li class="arrow"><a href="#contactus">Contact Us</a></li>

        <li class="arrow"><a href="#createacct">create account</a></li>

    </ul>

</div>
```

Let's run the application to see how it works now. When we run the program, we'll see that the *home* panel now has three list items instead of two, as show in Figure 2.7(a). When we select the *Create Account* item, we will jump to the *Create Account* panel, as shown in Figure 2.7(b). Each text field (*Userid, Contact No, and Email Address)* contains placeholder text to help the user complete the field entries. Figure 2.7(c) shows the form after user information has been entered.

Figure 2.7. (a) *Home* panel with three list items (b) Form with three input text fields (c) Form with user's information entered

37

In the next chapter, we'll learn how to save user information in database tables. The next step, though, is learning to display user information and apply validation checks.

2.8. Displaying User Information

To display the information entered by the user, we need to do two things:

- Add a panel to the application that displays user-entered information
- Add a JavaScript function to the application that's invoked when the *Create Account* button is selected in the form. This function then retrieves the user information and displays it in the newly added panel

The complete code of the application after adding the new panel and JavaScript function is shown in Listing 2.3. The additions discussed here are shown in bold.

Listing 2.3. Application to Display and Retrieve User Information

```html
<html>
    <head>
        <title>Book Store</title>
        <link type="text/css" rel="stylesheet" media="screen" href="jqtouch/jqtouch.css">
        <link type="text/css" rel="stylesheet" media="screen"
            href="themes/apple/theme.css">
        <script type="text/javascript" src="jqtouch/jquery.1.3.2.min.js"></script>
        <script type="text/javascript" src="jqtouch/jqtouch.js"></script>
        <script src="extensions/jqt.autotitles.js" type="application/x-javascript"
            charset="utf-8"></script>
        <script type="text/javascript">
            var jQT = new $.jQTouch();
            $(function(){
                $('#newuser .whiteButton').click(function(){
                    var uid = $('#Userid').val();
                    var contact = $('#ContactNo').val();
                    var emailid = $('#EmailAddress').val();
                    $('#userdetails').append('<p> Name : '+uid+'</p>');
                    $('#userdetails').append('<p> Contact Number : '+contact+'</p>');
                    $('#userdetails').append('<p> Email Address : '+emailid+'</p>');
                    jQT.goTo('#displayinfo');
                    return false;
                });
            });
        </script>
    </head>
```

```
<body>
    <div id="home">
        <div class="toolbar">
            <h1>Book Store</h1>
        </div>
        <ul class="rounded">
            <li class="arrow"><a href="#books">Books</a></li>
            <li class="arrow"><a href="#contactus">Contact Us</a></li>
            <li class="arrow"><a href="#createacct">Create Account</a></li>
        </ul>
    </div>
    <div id="books">
        <div class="toolbar">
            <h1>Categories</h1>
            <a class="button back" href="#">Back</a>
        </div>
        <ul class="rounded">
            <li class="arrow"><a  href="#book">Literature & Fiction</a></li>
            <li class="arrow"><a  href="#book">Home & Garden</a></li>
            <li class="arrow"><a  href="#book">Computers & Internet</a></li>
            <li class="arrow"><a  href="#book">Cooking, Food & Wine</a></li>
        </ul>
    </div>
    <div id="book">
        <div class="toolbar">
            <h1>Books</h1>
            <a class="button back" href="#">Back</a>
        </div>
        <div class="info">
            <p>Following is the list of books available in this category</p>
        </div>
    </div>
    <div id="contactus">
        <div class="toolbar">
            <h1>Contact Us</h1>
            <a class="button back" href="#">Back</a>
        </div>
        <div class="info">
```

```
                    <p>XYZ Book Company</p>
                    <p>11 Books Street, NY, NY 10012 </p>
                    <p>USA</p>
                     Email us: <a href="mailto:bmharwani@yahoo.com"
                         target="_blank">bmharwani@yahoo.com</a>
                </div>
            </div>
            <div id="createacct">
                <div class="toolbar">
                    <h1>Create Account</h1>
                    <a class="button cancel" href="#">Cancel</a>
                </div>
                <form id="newuser">
                    <ul class="rounded">
                        <li><input type="text" id="Userid" placeholder="Userid" /></li>
                        <li><input type="text" id="ContactNo" placeholder="Contact No" /></li>
<li><input type="text" id="EmailAddress" placeholder="Email Address"/>
                        </li>
                    </ul>
                    <a href="#" class="whiteButton">Create Account</a>
                </form>
            </div>
            <div id="displayinfo">
                <div class="toolbar">
                    <h1>User Details</h1>
                    <a class="button back" href="#home">Home</a>
                </div>
                <p>User information is as follows :</p>
                <div id="userdetails">
                </div>
            </div>
        </body>
</html>
```

When the *whiteButton* button containing the text *Create Account* is selected, we jump to the *displayinfo* panel. The *displayinfo* panel contains a toolbar with the title *User Details*. This toolbar has two buttons: *button* and *back.* The *displayinfo* panel contains a paragraph element that displays a *User information is as follows:* message. You might be surprised by the empty *userdetails* div element that follows the paragraph element. This empty div is added to the panel so the user-entered information can be displayed through it.

The function *('#newuser .whiteButton').click(function()* is invoked when the user selects the *Create Account* button after entering information in the *newuser* form fields. The function retrieves the data entered in

Userid, ContactNo, and *EmailAddress* fields through their id's and stores it in the three variables *uid, contact,* and *emailid* respectively. Afterwards, the information contained in *uid, contact,* and *emailid* is wrapped inside the paragraph elements and appended to the *userdetails* div element.

The *userdetails* div element is a member of the *displayinfo* panel we added to the application to display user information. Once the information is appended to *userdetails*, we jump to the *displayinfo* panel and see the user data.

Let's execute the application. When the *Create Account* item in the *home* panel is selected, we jump to the *Create Account* panel shown in Figure 2.8(a). Figure 2.8(b) shows some sample user information. When the form is filled and the *Create Account* button selected, we jump to the *displayinfo* panel, where we see the user information, as shown in Figure 2.8(c).

| (a) | (b) | (c) |

Figure 2.8. (a) Form with three input text fields (b) Form with user's information entered

(c) User information displayed in the *User Details* panel

We've successfully displayed user information, but what if the user forgets to enter some essential data? To avoid this problem, we need to apply validation checks that will test the data and alert the user if an essential field is left blank.

2.9. Applying Validation Checks

Should a user goof during data entry, we want to give him or her a chance to go back and supply the missed information. We've seen that the JavaScript function *('newuser .whiteButton').click(function()* is invoked when the user selects the *Create Account* button in the form. Using this function, we will check the contents of the input text fields. We know by now that the function retrieves the data entered in the *Userid, ContactNo,* and *EmailAddress* text fields, *and* stores them in the three variables *uid, contact,* and *emailid* respectively. Let's modify the function, as shown below, to implement validity checks:

```
$(function(){
    $('#newuser .whiteButton').click(function(){
        var uid = $('#Userid').val();
        var contact = $('#ContactNo').val();
        var emailid = $('#EmailAddress').val();
        $('#userdetails p').remove();
        if(uid.length <=0)
        {
            $('#userdetails').append('<p> User id cannot be blank. Please supply userid
                </p>');
            jQT.goTo('#displayinfo');
            return;
        }
        if(contact.length <=0)
        {
            $('#userdetails').append('<p> Contact Number cannot be blank. Please supply
                contact number </p>');
            jQT.goTo('#displayinfo');
            return;
        }
        if(emailid.length <=0)
        {
            $('#userdetails').append('<p> Email Address cannot be blank. Please supply
                email address </p>');
            jQT.goTo('#displayinfo');
            return;
        }
        $('#userdetails').append('<p>User information is as follows :</p>');
        $('#userdetails').append('<p> Name : '+uid+'</p>');
        $('#userdetails').append('<p> Contact Number : '+contact+'</p>');
        $('#userdetails').append('<p> Email Address : '+emailid+'</p>');
        jQT.goTo('#displayinfo');
        return false;
    });
});
```

The sequence of events is as follows:

1. We first remove any previous message displayed via any paragraph element in the *userdetails* div element.

2. We check the length of data stored in the *uid, contact,* and *emailid* variables to determine if any of them is less than, or equal to, 0. An empty text field will certainly result in a string of zero length.

3. If we find a null string, we display an error message by wrapping it inside a paragraph element and appending it to the *userdetails* div element. Recall that the *userdetails* div element is part of the *displayinfo* panel used to show user information.

4. After appending the message to the *userdetails* div element, we go to the *displayinfo* panel, so that an error message can be displayed.

5. If none of the fields is empty, the *userdetails* div element displays the user id, email address, and contact number.

We shall also make one additional change to the application. In the *displayinfo* div element, we'll have the user navigate to the *createacct panel* instead of the *home* panel via the *back* button, so he or she can correct any mistakes. The modified *displayinfo* div element is shown here in bold.

```
<div id="displayinfo">

    <div class="toolbar">

        <h1>User Details</h1>

        <a class="button back" href="#">Back</a>

    </div>

    <div id="userdetails">

    </div>

</div>
```

Let's run the application to see the validation checks at work. Select the *Create Account* list item in the *home* panel and leave the *userid* field blank. Tap the *Create Account* button. The message: *User id cannot be blank. Please supply userid* is displayed, as shown in Figure 2.9. Then select the *back* button to return to the form and enter a user id. The corrected user details are then displayed.

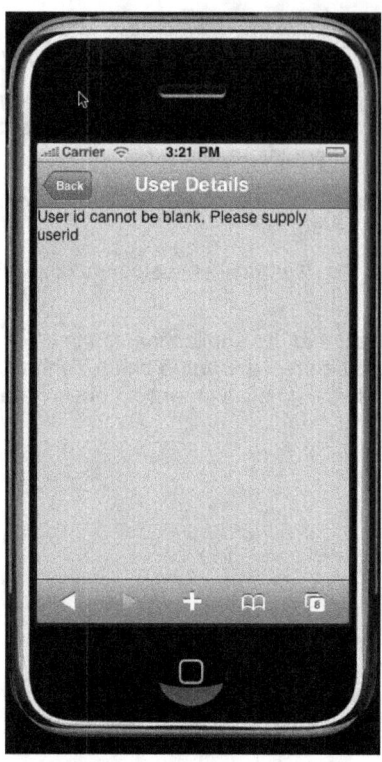

Figure 2.9. Error message displayed when the *userid* is left blank

2.10. Initializing jQTouch

So far, we've used default values for much of the application code. Let's now see how jQTouch is initialized and how it can be used to customize the appearance and behavior of our program. jQTouch is initialized with the following function:

```
var jQT = new $.jQTouch();
```

While initializing the jQTouch with this function, we can pass different property values to it to customize the the application's behavior. These property values are optional; Table 2.2 describes these properties.

Table 2.2. Properties That Can Be Passed During jQTouch Initialization

Property	Usage
fullScreen:	Makes the application full-screen when saved to the user's home screen. Default value is *true*. Set to *false* to disable.
Icon:	Sets the application's home screen icon. The icon can be a 57x57 pixel png image and is displayed when application is saved to the user's home screen. Default value is *null*.
addGlossToIcon:	Applies a glossy button effect to icon. Default value is *true*.
startupScreen:	Specifies the path to a 320x460 pixel startup screen. Applies to full-screen apps only.
statusBar:	Applies a style to the 20-pixel status bar when running as a full-screen app. Options are default, black, and black-translucent. The default value is *default*.

44

preloadImages:	Specifies an array of image paths we want loaded before the page loads.
useAnimations:	Enables or disables all animations. Default value is *true*
cacheGetRequests:	Caches GET requests, so that future requests for loaded data can be dealt with. Default value is *true*.
backSelector	Defines elements that will trigger the *back* behavior of jQTouch when tapped. The default is *back*.
cubeSelector	Defines elements that will trigger a cube animation from the current panel to the target panel.
dissolveSelector	Defines elements that will trigger a dissolve animation from the current panel to the target panel.
fadeSelector	Defines elements that will trigger a fade animation from the current panel to the target panel.
flipSelector	Defines elements that will trigger a flip animation from the current panel to the target panel. The default is *flip*.
fixedViewport	If set to true, this property will not allow zoom in or out on the page. The default is *true*.
formSelector	Defines elements that should be styled as a form by the CSS theme.
fullScreenClass	Defines the class name applied to the body when the application is launched in full-screen mode. The default is *true*.
popSelector	Defines elements that will trigger a pop animation from the current panel to the target panel.
slideInSelector	Defines elements that will trigger a slide-left animation from the current panel to the target panel. The default is *ul li a*.
slideUpSelector	Defines elements that will cause the target panel to slide up in front of the current panel. The default is *slideup*.
submitSelector	Defines selector that, when clicked, will submit the form and close the keyboard if it's open
swapSelector	Defines elements that will cause the target panel to swap into view in front of the current panel.

Here's an example of these properties at work:

```
var jQT = new $.jQTouch({
    icon         : 'bookIcon.png',
    addGlossToIcon : false,
    startupScreen : 'startSplash.png',
    statusBar    : 'black',
    preloadImages : [
                    'themes/apple/img/backButton.png',
                    'themes/apple/img/blueButton.png',
                    'themes/apple/img/cancel.png',
                    'themes/apple/img/grayButton.png',
                    'themes/apple/img/whiteButton.png',
```

```
                      'themes/apple/img/loading.gif'
        ]
});
```

When this code is added to Listing 2.3, the home screen icon is set to *bookIcon.png*, which will be displayed when the application is saved to user's home screen. No glossy button effect will be applied to the icon.

The startup screen, *startSplash.png*, will be displayed if the application runs as a full-screen app. The status bar will be black if the application runs as full-screen app, and six images (*backButton.png*, *blueButton.png*, *cancel.png*, *grayButton.png*, *whiteButton.png*, and *loading.gif*) will be loaded before the page loads.

   Note: By default the application runs as a full-screen app.

Let's finish the chapter by adding an icon to the home screen.

2.10.1. Adding an Icon to the Home Screen

Let's assume that we have a 57x57 pixel png file called *bookIcon.png* located in our application folder. The first thing we need to do is tap the circled *plus* button at the bottom of the Safari window, as shown in Figure 2.10.

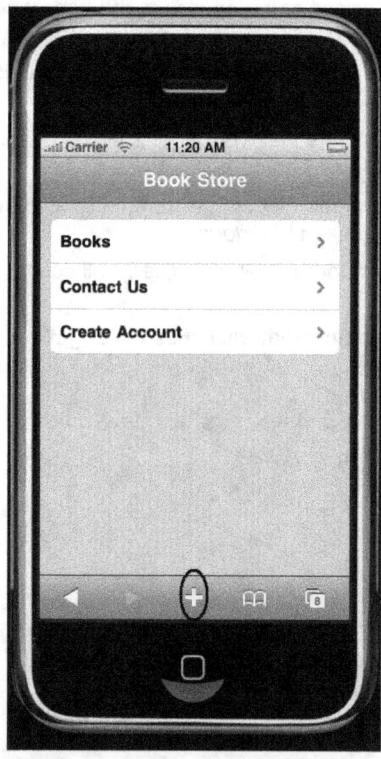

Figure 2.10. The circled *plus* button at the bottom of the Safari window

Three options will pop up on the screen, as shown in Figure 2.11. Select the *Add to Home Screen* button.

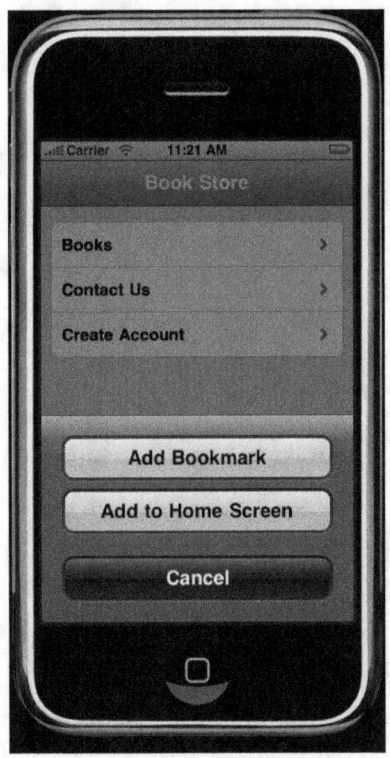

Figure 2.11. Dialog box showing the three options that appear after selecting the *plus* button from the Safari window

After selecting the *Add to Home Screen* button, we'll see the *Add to Home* panel, as shown in Figure 2.12. By default, the title *Book Store,* retrieved from the title of the *home* panel, will appear on the screen. We can change the title if we wish. Finally, select the *Add* button. By default, the iPhone applies rounded corners and a glossy effect to the icon, but in this case, we set the value of the *addGlossToIcon* property to *false*, so no glossy effect will be applied.

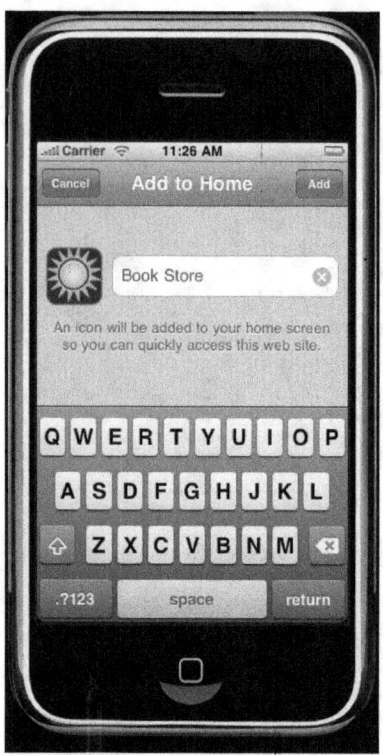

Figure 2.12. *Add to Home* **panel with the icon defined by the** *bookIcon.png* **image file**

The icon will now appear on the home screen. When tapped, the splash screen (from the *startSpash.png* image file) will display for a moment while the application loads,. as shown in Figure 2.13.

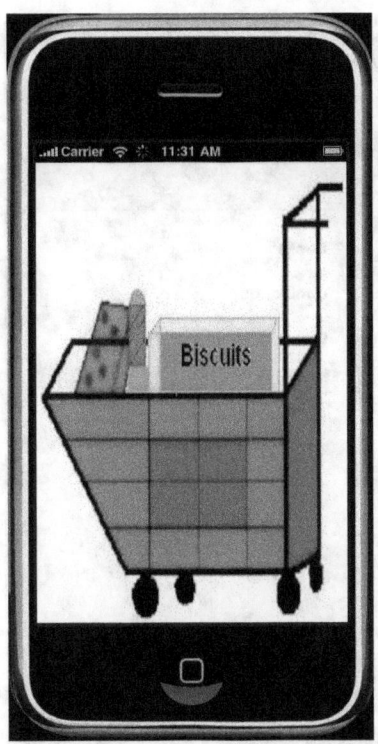

Figure 2.13. Splash screen, defined by the *startSplash.png* image file, appears while the application is loading

The splash screen will be followed by the *home* panel of our web application, as shown in Figure 2.14. Notice that, because the program is running as full-screen app, the background color of the 20-pixel status bar at the top of the screen is black, as we have set the *statusBar* property to *black*. The default color of the status bar is gray. We can also set its color to *black-translucent*, which makes it partially transparent.

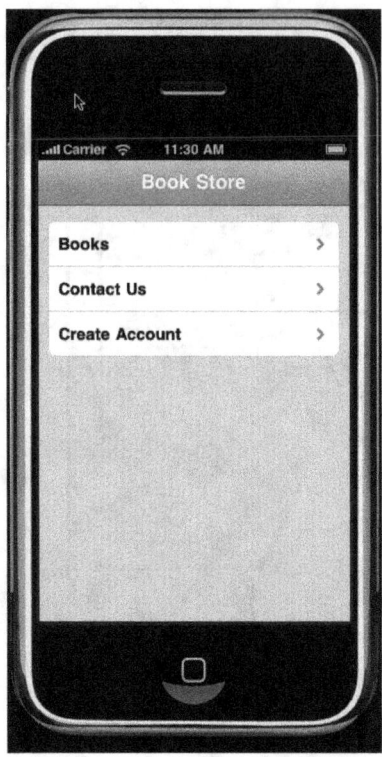

Figure 2.14. Home panel of the web application with a black status bar

2.11. Summary

In this chapter, we introduced jQTouch and used it to develop a small mobile application. We also learnt how to:

- Improve the application by adding additional panels

- Automatically display target titles, depending on the option selected in the previous view

- Highlight important information

- Add a *Create Account* form to the application

- Display information entered by the user and perform validation checks on the form

In the next chapter, we'll learn different ways of navigating between screens. We'll also see how to implement navigation when the Ajax technique is applied to the application.

3

Understanding Navigation

In the previous chapter, we talked about the role of the *Back* and *Cancel* buttons and how they return the display to a previous panel (also called a page or screen), using reverse animation. When we navigate to a panel, it's stored in the page history stack and, when a *Back* or *Cancel* button is selected, we jump to the top panel in the page history stack. In the previous chapter, we also saw that for forward navigation, we used an unordered list. Each list item contains a hyperlink that jumps to the appropriate panel.

In this chapter, we'll cover:

- Forward navigation via toolbar buttons

- Reverse navigation via the *Back* button

- Forward and reverse navigation via Panel buttons

- Navigation via the *goTo()* function

- Navigation via the *goBack()* function

- Working with Forms

- Introduction to PHP programming

- Collecting information sent by forms using $_GET, $_POST and $_REQUEST arrays

3.1 Forward Navigation Via Toolbar Buttons

In the application we created in Chapter 2, we used hyperlinks wrapped inside list items for forward navigation. Not surprisingly, we can use the toolbar buttons to do the same thing. To understand how this function works, let's create a small application with six panels: *Home, Categories, Subcategories, Select Books, Items in Cart* and *Checking Out*. Let's assume that the ids assigned to the six panels are *home, bookscategories, subcategories, booksdisplay, showcart,* and *checkout* respectively. The purpose of each panels is as follows :

1. **Home**—The first panel (id*home*) that appears on execution of application, as shown in Figure 3.1(a).

2. **Categories**—This panel (id *bookscategories*) displays book categories when the *Books* list item is selected from the *Home* panel, as shown in Figure 3.1(b).

3. **Subcategories**—This panel (id *subcategories*) displays book subcategories when a book category is selected from the *Categories* panel, as shown in Figure 3.1(c).

4. **Select Books**—This panel (id *booksdisplay*) displays the books belonging to the selected category and subcategory. The panel appears when a subcategory from *Subcategories* panel is selected. From this panel, the user can select books to be added to the cart, as shown in Figure 3.2(a).

   Note: The list items that appear in *Categories, Subcategories,* and the *Select Books* panels are hard-coded and just dummies for the time being. In later chapters, we'll learn how to fetch book categories from the *books* table and display the related subcategory. We'll also see how to display the actual book titles belonging to the selected category and subcategory.

5. **Items in Cart**—This panel (id *showcart*) displays the books in the cart. For the time being, only dummy books will be in the cart. This panel has a *Checkout button*, which jumps to the *Checking Out* panel, as shown in Figure 3.2(b).

   Note: For now, cart contents are hard-coded. We will discuss the actual code for adding, updating, and deleting books in future chapters.

6. **Checking Out**—This is the last panel of the *checkout* process and only has a *Back* button for returning to the previous screen, as shown in Figure 3.2(c).

Figure 3.1. (a) *Home* panel with *Show Cart* button (b) *Categories* panel showing book categories

(c) *Subcategories* panel showing book subcategories

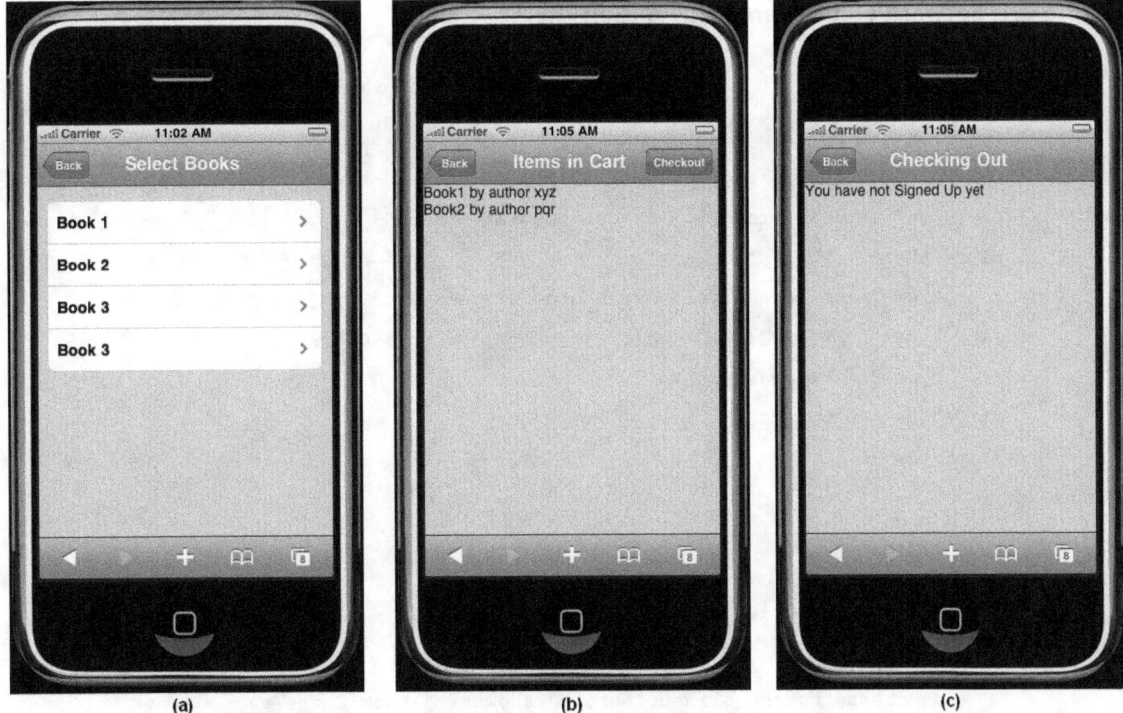

Figure 3.2. (a) *Select Books* panel showing books in the selected category and subcategory

(b) Items in the *Cart* panel, and the *Back* and *Checkout* buttons (c) *Checking Out* panel

Having built a similar application in Chapter 2, we can easily code the pages shown in Figures 3.1 and 3.2. We can create the *Back* buttons in all the screen panels by adding hyperlinks to the toolbars with the class name *back*. These href hyperlinks all point to #, which simply jumps to the first previous page in the in page history.

The only problem we face is making the two *Show Cart* and *Checkout* buttons work. But before we can deal with actual functionality, we first have to learn how to put buttons on the right side of the toolbar.

The *back* and *cancel* buttons appear on the left side of the toolbar. Only buttons with the *button* class name appear on the right side. To make the button jump to a panel with a specific id, we need to set the button *href* to point to it. You might think, at this point, that the code for a right-side button that navigates to the *showcart* id would look like this:

```
<a class="button " href="#showcart">Show Cart</a>
```

Although the code appears to be correct, it won't actually do what we want it to, because no animation has been specified in the hyperlink. The default *slide* animation won't work with this code fragment. It's a simple matter, though, to specify any of the other animation effects that will work: *slideup, dissolve, fade, flip, pop, swap, and cube*.

Functional code for the *showcart* id, using the *slideup* animation is:

```
<a class="button slideup" href="#showcart">Show Cart</a>
```

Similarly, the following fragment creates a *Checkout* button on the right side of the toolbar that jumps to the *checkout* panel with a *flip* animation:

```
<a class="button flip" href="#checkout">Checkout</a>
```

The complete code is shown in Listing 3.1.

Listing 3.1. Forward and Reverse Navigation

```html
<html>
    <head>
        <title>Book Store</title>
        <link type="text/css" rel="stylesheet" media="screen" href="jqtouch/jqtouch.css">
        <link type="text/css" rel="stylesheet" media="screen"
            href="themes/apple/theme.css">
        <script type="text/javascript" src="jqtouch/jquery.1.3.2.min.js"></script>
        <script type="text/javascript" src="jqtouch/jqtouch.js"></script>
        <script type="text/javascript">
            var jQT = new $.jQTouch();
        </script>
    </head>
    <body>
        <div id="home">
            <div class="toolbar">
                <h1>Home</h1>
                <a class="button slideup" href="#showcart">Show Cart</a>
            </div>
            <ul class="rounded">
                <li class="arrow"><a href="#bookscategories">Books</a></li>
            </ul>
        </div>
        <div id="bookscategories">
            <div class="toolbar">
                <a class="back"  href="#">Home</a>
                <h1>Categories</h1>
            </div>
            <ul class="rounded">
                <li class="arrow"><a  href="#subcategories">Literature & Fiction</a></li>
                <li class="arrow"><a  href="#subcategories">Home & Garden</a></li>
                <li class="arrow"><a  href="#subcategories">Computers & Internet</a></li>
                <li class="arrow"><a  href="#subcategories">Cooking, Food & Wine</a></li>
            </ul>
        </div>
        <div id="subcategories">
          <div class="toolbar">
                <a class="back"  href="#">Back</a>
```

```html
            <h1>Subcategories</h1>
        </div>
        <ul class="rounded">
            <li class="arrow"><a href="#booksdisplay">Subcategory1</a></li>
            <li class="arrow"><a href="#booksdisplay">Subcategory2</a></li>
            <li class="arrow"><a href="#booksdisplay">Subcategory3</a></li>
        </ul>
    </div>
    <div id="booksdisplay">
        <div class="toolbar">
            <a class="back"  href="#" >Back</a>
            <h1>Select Books</h1>
        </div>
        <ul class="rounded">
            <li class="arrow"><a href="#showcart">Book 1</a></li>
            <li class="arrow"><a href="#showcart">Book 2</a></li>
            <li class="arrow"><a href="#showcart">Book 3</a></li>
            <li class="arrow"><a href="#showcart">Book 3</a></li>
        </ul>
    </div>
    <div id="showcart">
        <div class="toolbar">
            <a  class="back" href="#" >Back</a>
            <a class="button flip" href="#checkout">Checkout</a>
            <h1>Items in Cart</h1>
        </div>
        <p> Book1 by author xyz </p>
        <p> Book2 by author pqr </p>
    </div>
    <div id="checkout">
        <div class="toolbar">
            <a  class="back" href="#">Back</a>
            <h1>Checking Out</h1>
        </div>
        <p> You have not Signed Up yet </p>
    </div>
</body>
</html>
```

Note: Buttons of class *back* or *cancel* may only appear on the left side of a toolbar. Only buttons of class *button* may appear on the right.

Now let's see how the *Back* button can be set to navigate to any page in the page history, not just the immediate previous page.

3.2 Reverse Navigation Via the Back Button

In the example above, the *href* of each *Back* button was set to #, which navigates to the previous panel. To have a *Back* button navigate to any panel in the page history, we merely have to set its *href* to point to that panel instead. For example, if an *href* points to #home, tapping that button causes the app to display the *home* panel.

Let's implement following changes to our application:

- The *Back* button of the *Subcategories* panel (id *subcategories*) will be set to navigate to the *Home* panel (id *home*) instead of the preceding panel, *Categories*.

- The *Back* button of the *Items in Cart* panel (id *showcart*) will be set to navigate to the *Subcategories* panel instead of the preceding panel, *Select Books*.

The modified code of Listing 3.1 is shown in bold.

```
<div id="subcategories">

    <div class="toolbar">

        <a class="back"  href="#home">Home</a>

        <h1>Subcategories</h1>

    </div>

    <ul class="rounded">

        <li class="arrow"><a href="#booksdisplay">Subcategory1</a></li>

        <li class="arrow"><a href="#booksdisplay">Subcategory2</a></li>

        <li class="arrow"><a href="#booksdisplay">Subcategory3</a></li>

    </ul>

</div>

<div id="showcart">

    <div class="toolbar">

        <a  class="back" href="#subcategories" >Subcategories</a>

        <a class="button flip" href="#checkout">Checkout</a>

        <h1>Items in Cart</h1>

    </div>

    <p> Book1 by author xyz </p>

    <p> Book2 by author pqr </p>

</div>
```

The *href* of the *Home* button hyperlink now points to #home. When the *Home* button is selected in the *subcategories* toolbar, the *Home* page will be displayed. Similarly the *href* of the *Subcategories* button hyperlink now points to #subcategories. When the *Subcategories* button in the *showcart* toolbar is tapped, the *subcategories* page will be displayed.

Let's run the application to see what happens now. As usual, the first panel we see when the application is executed is the *Home* panel that was shown Figure 3.1(a). If we select the *Books* list item, we'll still see the *Categories* panel that was shown in Figure 3.1(b). And selecting a book category displays the *Subcategories* panel. As shown in Figure 3.3(a). Now we can see that the toolbar button says *Home* instead of *Back,* as was shown in Figure 3.1(c). Obviously, tapping the *Home* button takes us back to the *Home* panel instead of the *Categories* panel.

If we choose a subcategory, we'll see the *Select Books* panel shown in Figure 3.2(a). After we select a book from the *Select Books* panel, the *Items in Cart* page is displayed, as shown in Figure 3.3(b). And, we can see that the left-side toolbar button now points to *Subcategories* instead of *Back* (which returned us to the *Select Books* panel).). Tapping the *Subcategories* button takes us directly to the *Subcategories* panel instead of the preceding panel, *Select Books*.

Figure 3.3. (a) *Subcategories* panel with the *Back* button pointing to the *Home* panel

(b) *Items in Cart* panel with the *Back* button pointing to the *Subcategories* panel

As long as a page exists in the page history, we can set the *Back* button to point to it. If no such page exists in the history, by default, the *Back* button simply returns to the previously displayed panel.

3.3 Navigating using goTo() function

jQTouch provides a *goTo()* function that can be used to go to any page (panel) irrespective of whether the page exists in page history or not. The syntax for using the function is:

```
goTo('panel_id', 'animation effect');
```

where *panel_id* refers to the id of the panel we want displayed. The *animation effect* may be any of the eight types—*slide, slideup, dissolve, fade, flip, pop, swap,* and *cube*—we want applied to the panel jump. Recall that if we don't specify an animation effect, the default, *slide in,* is used. For example, this code will navigate the user to the *showcart* panel via the *slide* effect.

```
jQT.goTo('#showcart','slide');
```

Let's apply *goTo()* function to the code in Listing 3.1. In that example, the code to navigate to the *Show Cart* panel from the *Home* page was as follows:

```
<a class="button slideup" href="#showcart">Show Cart</a>
```

Remember that this little bit of code doesn't actually work, because the *slide* animation isn't supported here. So let's apply the *goTo*() function and replace the non-working fragment with this:

```
<a class="button" href="#" onclick="jQT.goTo('#showcart','slide');">Show Cart</a>
```

This line causes the program to jump from the *Home* page to the *showcart* panel using the *slide* animation.

Similarly, the following code in Listing 3.1:

```
<a class="button flip" href="#checkout">Checkout</a>
```

can be replaced by:

```
<a class="button" href="#" onclick="jQT.goTo('#checkout','slide');">Checkout</a>
```

This code line jumps from the *Showcart* panel to the *Checkout* page when the *Checkout* button is tapped.

We've covered the code for toolbar button navigation, so let's now let's take a look at buttons that appear elsewhere in a panel.

3.4 Forward and Reverse Navigation via Panel Buttons

In Chapter 2, we learned how to program forward navigation by wrapping hyperlinks inside of list items. In this section, we're going to take a different approach.

Let's say that we want to provide two navigational options on the *Checking Out* panel

- Move forward to the next panel to create an account

- Go back to the *Categories* panel to do more shopping

We want the *Checking Out* panel to have two buttons, *Create Account* and *More Shopping,* as shown in Figure 3.4(a). To add the two buttons in the *Checking Out* panel (id *checkout*), the code in Listing 3.1 will be modified as shown below:

```
<div id="checkout">

    <div class="toolbar">

        <a  class="back" href="#">Back</a>

        <h1>Checking Out</h1>

    </div>

    <p> You have not Signed Up yet </p><br/>

    <a href="#" id="createaccount" class="whiteButton">Create Account</a><br/>

    <a href="#" id="moreshopping" class="whiteButton">More Shopping</a>

</div>
```

A paragraph element has been added to the *Checking Out* panel informing the user that he or she hasn't yet signed up for an account. Two hyperlinks follow the paragraph element: *createaccount* and *moreshopping.* These hyperlinks are assigned to the class name *whiteButton,* which will make them look (surprise!) like buttons. To make these buttons operational, we need to do two things:

- Add a *Create Account* panel with the appropriate input fields for creating an account
- Add two JavaScript functions (one for each button) that senses the click event and navigates to the appropriate panel.

To add a *Create Account* panel, add the following code to Listing 3.1:

```
<div id="createacct">
    <div class="toolbar">
        <h1>Create Account</h1>
        <a class="button cancel" href="#">Cancel</a>
    </div>
    <form id="newuser">
        <ul class="rounded">
            <li><input type="text" class="Userid" id="Userid" placeholder="Userid"/></li>
            <li><input type="text" id="ContactNo" placeholder="Contact No" /></li>
            <li><input type="text" id="EmailAddress" placeholder="Email Address"/></li>
        </ul>
        <a href="#" class="whiteButton">Create Account</a>
    </form>
</div>
```

The *Create Account* panel is assigned the id *createacct* and a button in the toolbar is added with class names *button* and *cancel. When* this button is selected, the account creation operation is cancelled and the user is returned to the *Checking Out* panel.

The new panel contains a form with a *newuser* id and three text input fields wrapped inside list items. The three text fields contain placeholder text, *Userid, Contact No* and *Email Address* as an input guide. At the end of the form, a *Create Account* button is created . At the moment, the *Create Account* button does absolutely nothing, but in the later sections of this chapter we'll see how tapping the this button leads to a PHP script for further processing. We'll also learn how data entered in the input fields are stored in server-side database tables.

To make the *Create Account* and the *More Shopping* buttons operational we need to add the two following JavaScript functions to the code in Listing 3.1:

```
$('#createaccount').click(function(){
    jQT.goTo('#createacct','slide');
});
$('#moreshopping').click(function(){
    jQT.goTo('#bookscategories','slide');
});
```

The first JavaScript function is invoked when a click event occurs on the *createaccount* button. The JavaScript contains a *goTo()* function that navigates us to the *Create Account* panel using a *slide* animation. The second JavaScript function is invoked when a click event occurs on the *More Shopping* button and jumps to the *bookscategories* panel using a *slide* animation effect.

The revised application is shown in Listing 3.2, with new code shown in bold..

Listing 3.2. Forward and Reverse Navigation Using Panel Buttons

```html
<html>
    <head>
        <title>Book Store</title>
        <link type="text/css" rel="stylesheet" media="screen" href="jqtouch/jqtouch.css">
        <link type="text/css" rel="stylesheet" media="screen"
            href="themes/apple/theme.css">
        <script type="text/javascript" src="jqtouch/jquery.1.3.2.min.js"></script>
        <script type="text/javascript" src="jqtouch/jqtouch.js"></script>
        <script type="text/javascript">
            var jQT = new $.jQTouch();
            $(function(){
                $('#createaccount').click(function(){
                    jQT.goTo('#createacct','slide');
                });
                $('#moreshopping').click(function(){
                    jQT.goTo('#bookscategories','slide');
                });
            });
        </script>
    </head>
    <body>
    <div id="home">
        <div class="toolbar">
            <h1>Home</h1>
            <a class="button" href="#" onclick="jQT.goTo('#showcart','slide');" >Show
                Cart</a>
        </div>
        <ul class="rounded">
            <li class="arrow"><a href="#bookscategories">Books</a></li>
        </ul>
    </div>
    <div id="bookscategories">
        <div class="toolbar">
            <a class="back" href="#">Home</a>
            <h1>Categories</h1>
        </div>
        <ul class="rounded">
```

```
            <li class="arrow"><a  href="#subcategories">Literature & Fiction</a></li>
            <li class="arrow"><a  href="#subcategories">Home & Garden</a></li>
            <li class="arrow"><a  href="#subcategories">Computers & Internet</a></li>
            <li class="arrow"><a  href="#subcategories">Cooking, Food & Wine</a></li>
        </ul>
    </div>
    <div id="subcategories">
        <div class="toolbar">
            <a class="back"  href="#">Back</a>
            <h1>Subcategories</h1>
        </div>
        <ul class="rounded">
            <li class="arrow"><a href="#booksdisplay">Subcategory1</a></li>
            <li class="arrow"><a href="#booksdisplay">Subcategory2</a></li>
            <li class="arrow"><a href="#booksdisplay">Subcategory3</a></li>
        </ul>
    </div>
    <div id="booksdisplay">
        <div class="toolbar">
            <a class="back"  href="#" >Back</a>
            <h1>Select Books</h1>
        </div>
        <ul class="rounded">
            <li class="arrow"><a href="#showcart">Book 1</a></li>
            <li class="arrow"><a href="#showcart">Book 2</a></li>
            <li class="arrow"><a href="#showcart">Book 3</a></li>
            <li class="arrow"><a href="#showcart">Book 3</a></li>
        </ul>
    </div>
    <div id="showcart">
        <div class="toolbar">
            <a  class="back" href="#" >Back</a>
            <a class="button" href="#" onclick="jQT.goTo('#checkout','slide');">
                Checkout</a>
            <h1>Items in Cart</h1>
        </div>
        <p> Book1 by author xyz </p>
        <p> Book2 by author pqr </p>
```

```
        </div>

    <div id="checkout">

        <div class="toolbar">

            <a   class="back" href="#">Back</a>

            <h1>Checking Out</h1>

        </div>

        <p> You have not Signed Up yet </p><br/>

        <a href="#createacct" id="createaccount" class="whiteButton">Create

            Account</a><br/>

        <a href="#" id="moreshopping" class="whiteButton">More Shopping</a>

    </div>

    <div id="createacct">

        <div class="toolbar">

            <h1>Create Account</h1>

            <a class="button cancel" href="#">Cancel</a>

        </div>

        <form id="newuser">

            <ul class="rounded">

                <li><input type="text" class="Userid" id="Userid" placeholder="Userid"/>

                    </li>

                <li><input type="text" id="ContactNo" placeholder="Contact No" /></li>

                <li><input type="text" id="EmailAddress" placeholder="Email

                    Address"/></li>

            </ul>

            <a href="#" class="whiteButton">Create Account</a>

        </form>

    </div>

</body>

</html>
```

Let's run the application and see what happens. The panels titled *Home, Categories, Subcategories, Select Books* and *Items in Cart* are unchanged.

When the *Checkout* button in the *Items in Cart* toolbar is selected, the *Checking Out* panel appears with two new buttons, as shown in Figure 3.4(a). If the user selects the *Create Account* button, *Create Account* panel is displayed, as shown in Figure 3.4(b). This panel has three input fields with placeholder text in gray color to guide what has to be entered in the respective input fields. The toolbar *Cancel* button jumps back to the *Checking Out* panel. When the *More Shopping* button in this panel is tapped, the *Categories* panel, shown in Figure 3.4(c), is displayed.

Note: If the user selects the *More Shopping* button from the *Checkout* panel without first having visited the *Categories* panel, the user is returned to the *Home* panel, because the *Categories* panel doesn't exist in the page history.

(a)   (b)   (c)

Figure 3.4. (a) *Checking Out* panel with *Create Account* and *More Shopping* buttons **(b)** *Create Account* panel with a *Cancel* button **(c)** *Categories* panel with a *Back* button pointing to the *Home* panel

3.4.1 Using goBack() function

To jump to any panel that exists in the page history, we can use the *goBack* function. The syntax of this function is:

```
goBack('panel_id/ number_of_pages', 'animation effect');
```

- *panel_id* refers to the panel id in the page history we want to jump to.
- *number_of_pages* designates how many prior pages in the page history we want to jump to. For example, if *number_of_pages* is set to 2, we jump back two pages in the page history.
- *animation effect* can be any of the eight animations we want to apply to the transition. If we don't specify an animation effect, the default *slide* animation slide is applied.

This bit of code returns us to the *bookscategories* panel:

```
jQT.goBack('#bookscategories','slide');
```

And this one jumps four pages back in the page history:

```
jQT.goBack(4,'slide');
```

We can replace the following JavaScript code in Listing 3.2 by any of the following three code fragments to produce the output shown in Figure 3.4:

```
$('#moreshopping').click(function(){

   jQT.goTo('#bookscategories','slide');

});
```

or

```
$('#moreshopping').click(function(){
    jQT.goBack('#bookscategories','slide');
});
```

or

```
$('#moreshopping').click(function(){
    jQT.goBack(4,'slide');
});
```

3.5 Working with Forms

Forms are indispensible for acquiring user feedback or information. A form consists of input fields, such as text, checkboxes, radio buttons. The data entered in these fields is passed on to a specified script for further processing after the *Submit* button has been clicked. The script can be any appropriate language,for example, PHP, ASP.NET, or Python. In this book, we'll be using PHP as the server-side scripting language.

Here's an example of a form:

```
<form action="createuser.php" method="POST" class="form">
    <ul class="rounded">
        <li><input type="text" class="Userid" name="Userid" placeholder="Userid" /></li>
        <li><input type="text" name="ContactNo" placeholder="Contact No"/></li>
        <li><input type="text" name="EmailAddress" placeholder="Email Address"/> </li>
    </ul>
    <a href="#" class="submit whiteButton">Submit</a>
</form>
```

Input text fields are assigned unique names—*Userid, ContactNo,* and *EmailAddress*—because data entered in the input fields is accessed by name only.

The attribute *action* points to a PHP file called *createuser.php*. The attribute *method* is set to *POST,* which means that the data entered into the input fields will be submitted to the *createuser.php* script using the HTTP request method POST for further processing. Form contents can be submitted to the specified script using either the GET or POST HTTP request methods.

- GET—This method of passing form information is considered less secure than POST, as the data is actually displayed in the browser's address bar, passed along with the URL of the script. For example, let's assume that the PHP script to which we want to pass form's information is *createuser.php,* and the names assigned to the input fields are *Userid, ContactNo,* and *EmailAddress*. When the Submit button is tapped, the browser's address bar would be:

    crreateuser.php? Userid=John123&ContactNo=123456789&EmailAddress=johny123@gmail.com

    Moreover, there is a limit on the amount of information that can be passed through the GET method.

- POST—This method is usually preferred because the information it is not displayed in the browser's address bar and can handle larger amounts of data.

Note: The information sent by a form through GET and POST methods is stored in the associative arrays $_GET and $_POST respectively.

Before we learn how to collect form information through a PHP script, let's talk briefly about the PHP language.

3.6 Introduction to PHP programming

A PHP script usually contains HTML tags and some embedded PHP scripting code. PHP commands are enclosed within special start and end tags: *<?php* and *?>*, so the interpreter can distinguish PHP code from HTML commands. Here's an example:

```
Syntax :

<?php

PHP commands

.................

.................


?>
```

The *<?php* tag tells the interpreter to open PHP mode. The *?>* tag tells the interpreter to close PHP mode.

The PHP code between *<?php* and *?>* is converted into HTML code by the PHP interpreter. The only data sent to the client's browser is the PHP code output (in HTML format), making it secure.

For example:

```
<html>

    <body>

        <?php echo "Welcome to our bookstore"; ?>

    </body>

</html>
```

This simple script shows that PHP commands can be easily embedded with HTML with the help of special start and end tags.

3.6.1 echo

The PHP *echo* command is used to output text to the browser.

For example:

1. echo 'Welcome to our bookstore';

2. $name="John";

    echo "Your name is: $name";

3. $a=10;

    $b=20;

    echo $a+$b;

The first example sends the bolded text, *Welcome to our bookstore*. The second example actually demonstrates how the variables are handled in PHP. The value to the variable is assigned with via the assignment operator **(=)**. All PHP variables start with a **$** sign symbol, followed by either a letter or underscore (no other character is allowed). A string, *John,* is assigned to the *$name* variable and is sent to the browser via the echo command. In the second example, the browser would display *Your name is John.* The third example would display the result of a simple numerical calculation.

This short PHP intro is sufficient for understanding how information sent by a form can be collected by a PHP script for further processing. We'll talk more about PHP in later chapters.

3.7 Collecting Information Sent by a Form

In this section, we'll learn to collect the information sent by a form through a PHP script. Recall that the two HTTP request methods by which information of a form can be passed to a target PHP script are GET and POST. We also noted that the information is then stored in the associative arrays, *$_GET* and *$_POST*. The target PHP uses these associative arrays to retrieve the form information, so let's talk a bit about them.

3.7.1 $_GET array

The *$_GET array* collects the values sent from a form using the HTTP GET method. When the user clicks the *Submit* button, all the information entered by the user in the form's input fields form is stored in $_GET. The associative array stores the information in the form of key/value pairs. The keys are the names assigned to the form's input text fields and the values are the data entered by the user in the respective keys. Thus, the data in the $_GET array can be accessed by specifying the key names. For example, information stored in the *Userid* key is accessed via $_GET["Userid"].

The target PHP script, *createuser.php,* can extract the data from a $_GET array using following code:

```php
<?php
    $uid =trim($_GET['Userid']);
    $contactno =trim($_GET['ContactNo']);
    $emailid =trim($_GET['EmailAddress']);
    echo 'Welcome ' . $uid . '!. Your account is created <br>';
    echo 'Your Contact number is ' . $contactno . '<br>';
    echo 'Your Email address  is ' . $emailid . '<br>';
?>
```

> Note: The period (.) in the *echo* command is used for concatenation. The *trim()* function truncates trailing spaces.

In this PHP script, we are using *$_GET* array to retrieve the information entered by the user in the form's input text fields *Userid, ContactNo,* and *EmailAddress,* and store them into the key variables *$uid, $contactno,* and *$emailid* respectively. The information stored in the key variables then displayed on the screen via the *echo* command.

3.7.2 $_POST array

The *$_POST* array collects the values sent from a form that is using the HTTP POST method. When the user clicks the Submit button, all the posted information will be stored in the $_POST associative array in the form of key/value pairs . The keys are the names assigned to the form's input text fields. Information in the *$_POST* array is accessed by specifying the respective key. For example, to access the information stored in the *Userid* input field, we would use $_POST["Userid"].

The target PHP script, *createuser.php,* extracts the data from $_POST array and displays a welcome message in this bit of code:

```php
<?php
    $uid =trim($_POST['Userid']);
```

```php
    $contactno =trim($_POST['ContactNo']);

    $emailid =trim($_POST['EmailAddress']);

    echo 'Welcome ' . $uid . '!. Your account is created <br>';

?>
```

3.7.3 $_REQUEST array

The $_REQUEST array contains the contents of both $_GET and $_POST. That is, the array is used to collect information from a form that is sending data by either the GET or POST method. If don't know which HTTP is being used by the form, it's a good idea to use a $_REQUEST array.

In the following code fragment, the target PHP script, *createuser.php*, extracts the data from a $_REQUEST array and displays a message to the user.

```php
<?php

    $uid =trim($_REQUEST['Userid']);

    $contactno =trim($_REQUEST['ContactNo']);

    $emailid =trim($_REQUEST['EmailAddress']);

    echo 'Welcome ' . $uid . '!. Your account is created <br>';

?>
```

3.8 Implementing Forms

We've covers forms, methods, and arrays, so we're finally ready to to process the data in the our application's *Create Account* panel.

The following fragment modifies the *Create Account* panel code shown in Listing 3.2. The code now takes user-entered data and sends it to a PHP script.

```html
<div id="createacct">

    <div class="toolbar">

        <h1>Create Account</h1>

        <a class="button cancel" href="#">Cancel</a>

    </div>

    <form action="createuser.php" method="POST" class="form">

        <ul class="rounded">

            <li><input type="text" class="Userid" name="Userid" placeholder="Userid"/>
                </li>

            <li><input type="text" name="ContactNo" placeholder="Contact No"/></li>

            <li><input type="text" name="EmailAddress" placeholder="Email Address"/></li>

        </ul>

        <a href="#" class="submit whiteButton">Submit</a>

    </form>

</div>
```

We can see that when the *Submit* button is pressed, we will go to the *createuser.php* file. The HTTP Request method used for passing the information entered in the input text fields is POST. That is, when the user fills out the form and clicks *Submit*, the $_POST["Userid"], $_POST["ContactNo"], and $_POST["EmailAddress"] elements will be automatically initialized with the data entered by the user in the input text fields named *Userid, ContactNo, and EmailAddress* respectively.

The information in the $_POST array can be retrieved by the *createuser.php* target script, which performs the following tasks:

- Extracts information from the $_POST array

- Creates a new panel for displaying extracted information

- Displays a message to the user

The script is shown below.

```php
<?php

    $uid =trim($_REQUEST['Userid']);

    $contactno =trim($_REQUEST['ContactNo']);

    $emailid =trim($_REQUEST['EmailAddress']);

?>

<div id="welcome">

    <div class="toolbar">

        <h1>Welcome</h1>

        <a class="back" href="#">Back</a>

    </div>

    <?php

        echo 'Welcome ' . $uid . '!. Your account is created ';

    ?>

</div>
```

Note: Remember that the $_REQUEST array can be used for extracting information sent by a form, regardless of the method (GET or POST) used by the form.

When a user enters information in the form and selects the *Submit* button, the data in the input text fields is passed to the *createuser.php* PHP script, as shown in Figure 3.5(a). We are using $_REQUEST array in the PHP script to retrieve the user-entered information, *Userid, ContactNo, and EmailAddress*, and store it into the variables *$uid, $contactno, and $emailid* respectively. The code also creates a *Welcome* panel containing a toolbar with a *Back* button and a welcome message, displayed in the panel body via the *echo* command, as shown in Figure 3.5(b).

Note: When user selects the *Submit* button, the application automatically jumps from the current panel to the target PHP script. We don't need to explicitly use a *goTo* function.

68

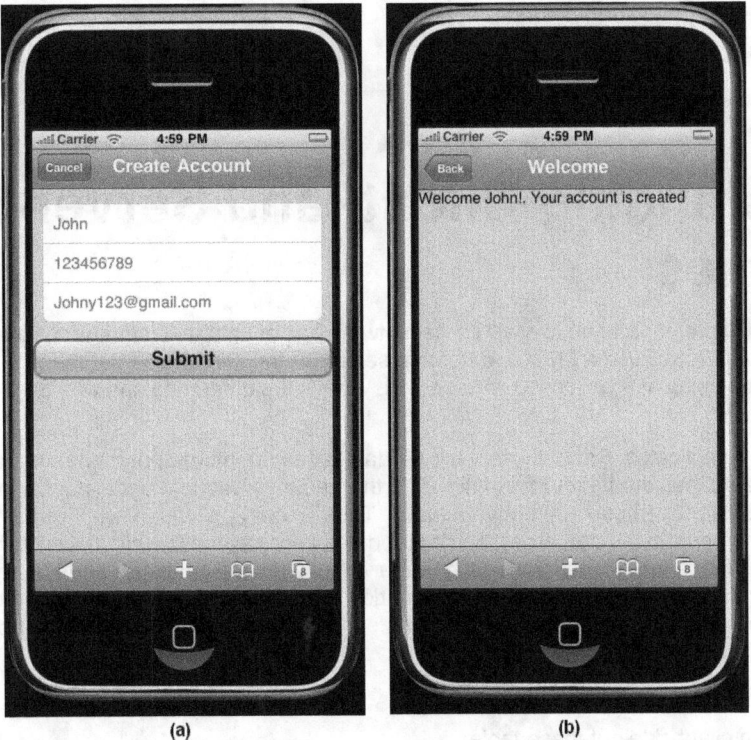

Figure 3.5. (a) Information entered in the form's input fields (b) *Welcome* panel with toolbar and account confirmation message

3.9 Summary

In this chapter, we examined the functions used for forward and reverse navigation. We also learned how to create toolbar and panel buttons. Our short PHP introduction showed how form information can be collected by a target PHP script for further processing. And we also learned how form information can be extracted from the three associative arrays $_GET, $_POST and $_REQUEST. In the next chapter we will learn about local and remote databases.

4

Understanding Client- and Server-Side Databases

In our sample *Bookstore* application, we will be using two types of database storage mechanisms: *Server-side* and *Client-side*. We will use a MySQL database server for the server-side database. For the client-side database, we will be using a JavaScript database API. What's the difference and why do we need both of these?

The answer is not complicated. Basically server-side databases are more appropriate for permanent data, while client-side databases are better for volatile data that requires frequent updating. Client-side data handling reduces server overhead and network traffic. Thus, in our application, we'll be using a server-side database for storing book, customer, order, and order detail information. The client-side database will be used to store cart information, which may change on a whim when the customer adds or deletes purchased items. Once a purchase has been completed, information from the cart table is transferred to the server-side database tables for long-term storage. Once the transfer is complete, the data is deleted from the client-side cart table.

In this chapter, we will learn about:

- Creating a client-side database table
- Inserting rows in a client-side table
- Updating a client-side table
- Fetching rows from a client-side table
- Accessing MySQL from PHP
- Fetching rows from a server-side database table
- Inserting rows in a server-side database table
- Querying a server-side database table
- Session management
- localStorage and sessionStorage

Let's begin with creating a client-side database and tables.

4.1 Creating a Client-Side Database and Tables

As we said in the introduction, we'll be using a JavaScript database API for the client-side database. This API provides a simple method of storing, retrieving, and marinating user-supplied information. In addition, it provides a relational database model, and so can easily handle complex data. We can use standard SQL statements to perform all database-related tasks, such as creating databases and tables, as well as inserting, fetching, deleting, and updating rows in the tables.

Let's start off by creating and opening a database. The code for creating and opening the client-side database is shown below:

```
var datab;
```

```
var shortName = 'tmpCart';

var version = '1.0';

var displayName = 'tmpCart';

var maxSize = 200000;

datab = openDatabase(shortName, version, displayName, maxSize);
```

The usage of the variables used in the code fragment is shown Table 4.1.

Table 4.1. Client-Side Database Variables

Variable	Usage
datab	Holds the reference to the database connection when it is established.
shortName	Stores the name of the database we will be creating on the client side.
version	Stores the version number assigned to the database. The version number is often needed for upgrades or changes to the database.
displayName	Stores the database name available to the user.
maxSize	Stores the expected size of our database in kilobytes. Should the size exceed the limit in this variable, we will be notified and asked if the increase should be allowed or denied.

The variable values defined in our code fragment say that we want to create a client-side database named *tmpCart* with a version number of 1.0 and a size limit of 200000 KB. After assigning these variable values. We can pass them to the *openDatabase* command, which actually creates the *tmpCart* database and stores the connection in a *datab* variable.

The code fragment can also be written as:

```
datab = openDatabase('tmpCart',  '1.0', 'tmpCart', 200000);
```

That is, we can directly specify the parameter values in the *openDatabase* function without using variables at all.

Now that we've created and opened a server-side database, let's create a table in it.

4.1.1 Creating Client-Side Tables

As we said earlier, we'll be using the client-side database for storing cart information. Let's first create a table called *shopcart* in our client-side database *tmpCart*, as shown in this code fragment:

```
datab.transaction(
    function(transaction) {
        transaction.executeSql(
            'CREATE TABLE  IF NOT EXISTS shopcart ' +
                ' (id INTEGER NOT NULL PRIMARY KEY AUTOINCREMENT, ' +
                ' cart_sess varchar(50), cart_isbn varchar(30),  cart_item_name
                varchar(100), cart_qty integer, cart_price float );'
        );
    }
);
```

The JavaScript database API supports SQL transactions and all database queries must take place in the context of a transaction. To execute a standard SQL query, we need to call the transaction object's *executeSql* method. To do this, we call the *transaction* method of the *datab* object and pass it an anonymous function. We then pass the transaction to the anonymous function so we can call the transaction object's *executeSql* method.

The *executeSql* method creates a table called *shopcart* if it doesn't already exist. The table has four fields: *id, cart_sess, cart_ISBN, cart_item_name, cart_qty,* and *cart_price.* The *id* field is the primary key. It's value cannot be null and is unique for each row of the table. We set this field to *AUTOINCREMENT*, so its value will increase by 1 for each table row added.

4.2 Inserting Rows in Client-Side Tables

Let's assume that a user has dropped a book in the cart, so we now have some cart data to process— session id, ISBN, title, quantity, and price. This data is stored in the variables *sid, isbn, title, qty,* and *price* respectively. We'll transfer this data to a row in the client-side *shopcart* table, as shown in the following code fragment.

```
datab.transaction(

    function(transaction) {

        transaction.executeSql(

            'INSERT INTO shopcart (cart_sess, cart_isbn,  cart_item_name, cart_qty,

                cart_price) VALUES (?,?,?,?,?);',

            [sid, isbn, title, qty, price],

            function(){

            },

            displayerrormessage

        );

    }

);
```

> Note: The Session id is used for session management—a mechanism used for tracking users visiting our web application. The Server generates a unique id for the current session. We'll look at this process in detail later in this chapter.

First we use the *transaction* object's *executeSql* method to execute a SQL query. We need to pass the data for five fields, or columns, to this method. The five question marks (?) in the VALUES parentheses are placeholders and take on the values in the array *[sid, ISBN, title, qty, price].* The first question mark is replaced by the value in the *sid* variable, the second question mark is replaced by the *ISBN* value, and so on.

The array of values is followed by an anonymous function called the *data callback* function, which may contain statements to execute after successful execution of the SQL statement. For example, *data callback* may contain calls to other functions that display a confirmation message or navigate to an other panel. If we don't want any action to take place after the successful execution of the SQL statement, we can leave this function empty, as we have actually done in the code fragment.

The last part of the fragment, *displayerrormessage,* is a call to the *error handling* function we want to evoke if the SQL statement fails. Here is an example of *displayerrormessage* usage:

```
function displayerrormessage(transaction, error) {

    alert('Error:  '+error.message+' has occurred with Code: '+error.code);
```

```
      return true;

}
```

Two parameters are passed to the *displayerrormessage* function: the *transaction* object and the *error* object. The *error* object displays the error message and the error code. The reason for passing *transaction* object to the *displayerrormessage* function is to allow more SQL statements to be executed from within the function, if desired. The *displayerrormessage* function may return *true* to halt the execution and roll back the entire transaction, or *false*, in which case the transaction will continue.

4.2.1 Error Codes

Table 4.2 contains a quick look at the common error codes generated while executing a transaction on client-side database tables.

Table 4.2 Error Codes and Occurrence

Error Code	Occurs When
0	The transaction failed for other non-database-related error.
1	The transaction failed for other database-related error.
2	The transaction failed because the version of the database didn't match the one user requested.
3	The transaction failed because the data returned from the database was too large.
4	The transaction failed because either there was not enough storage space left or the user didn't wanted the database to grow beyond the existing limit.
5	The transaction failed because the transaction included a syntax error, number of parameters mismatch, statement modifying the database in a read-only transaction, and so on.
6	The transaction failed because of constraint failure, for example, assigning two rows the same value in the primary field.

If we aren't interested in capturing errors, we can omit both *data callback and error handling callback* functions. If we do that, the code fragment will appear as follows:

```
datab.transaction(

    function(transaction) {

        transaction.executeSql(

            'INSERT INTO shopcart (cart_sess, cart_isbn,  cart_item_name, cart_qty,

                cart_price) VALUES (?,?,?,?,?);',

            [sid, isbn, title, qty, price]

        );

    }

);
```

4.3 Updating Client-Side Tables

Let's assume we want to change the quantity of a book with a particular ISBN in our *shopcart* table. To update the table, we'll need three pieces of data: the book's ISBN, the new quantity, and the session id.

Remember that every user of our application is assigned a unique session id by the server. The code for updating the client-side database is as follows:

```
datab.transaction(

    function(transaction) {

        transaction.executeSql(

            'update shopcart set cart_qty=? where cart_sess=? and cart_isbn=?;',

            [qty, sid, isbn]

        );

    }

);
```

We don't want to capture and display errors, so both the *data callback* and *error handling callback* functions are omitted. The code is quite simple. First the *executeSql* method is called to execute the SQL statement. We pass three placeholder parameters in the SQL UPDATE statement. As before, these parameters are sequentially replaced by the array of values that follow. In this case, the array consists of three variables, *[qty, sid, isbn]*.

4.4 Fetching Rows from Client-Side Tables

Every web application includes a technique known as *session management*, in which the server assigns a unique session id to every user visiting our web application. The session id helps to track users and is valid for the session, that is, until either the user exits the application or leaves the application idle for a specific time period. The session id is useful not only for recognizing a particular user, but also for tracking selected products or user-entered data.

We'll need the session id (*sid*) to display the information stored in the *shopcart* client-side database table. Assuming we have that information, the code for fetching rows of the *shopcart* table containing the user-selected book information is shown here:

```
datab.transaction(

    function(transaction) {

        transaction.executeSql(

            'SELECT cart_isbn,  cart_item_name, cart_qty, cart_price FROM shopcart where

                cart_sess=?;',[sid],

            function (transaction, result) {

                for (var i=0; i < result.rows.length; i++) {

                    var row = result.rows.item(i);

                    $('#showitems').append('<label>Isbn: </label>+row.cart_isbn+</em>');

                    $('#showitems').append('<label>Book: </label><em>' + row.cart_item_name

                        + '</em>');

                    $('#showitems').append('<label>Quantity: </label><em>' + row.cart_qty +

                        '</em>');

                    $('#showitems').append('<label>Price: </label>+row.cart_price+</em>');

                }

            },
```

```
        displayerrormessage

    );

  }

);
```

While executing a transaction that returns rows via the SQL SELECT query statement, we must use a *data callback* to handle the results. The *data callback* routine may contain code to perform tasks on the returned rows, such as displaying, processing, or saving to server-side database tables.

In the code fragment above, we execute the SQL statement as usual by calling *executeSql*. The SQL SELECT statement fetches rows from the *shopcart* table that matches the provided *sid* variable, and stores the rows in the *result* object. Thereafter, the role of *data callback* routine is to handle the returned rows.

4.4.1 Using a Data Callback Routine to Handle Returned Rows

In the *data callback* function following the SQL statement, we pass two parameters: *transaction* and *result*. The *transaction* parameter may be used to further call any *executeSql* method if we want to run more SQL statements. The *result* parameter is used to fetch rows collected in the *result* object on execution of the SQL statement. In code fragment above, we display the table rows by using a *for* loop that executes for each row fetched in the *result* object. During each loop, a single row is retrieved and temporarily stored in the variable *row*.

Thereafter, the data in each field is displayed via the *row* variable. The field data of each row are displayed by appending them to the *div* element of *showitems*. We are assuming that *showitems* actually exists in the code to display the rows of the table. The error-handling function, *displayerrormessage,* will be called if the SQL statement fails.

4.5 Accessing MySQL from PHP Functions

Now that we've covered how information is stored, retrieved, and updated in client-side database tables, let's take a look at similar functions in server-side databases. As we mentioned in the beginning of the chapter, we'll be using MySQL server for these tasks. Let's look at the functions we'll be using for selecting databases, accessing tables, executing SQL statements, and fetching results.

mysql_connect()

This function is used to establish the connection to the MySQL database server. It returns *true* if the connection is established, else *false* is returned. This function takes three parameters: *servername*, *userid*, and *password*.

Syntax

```
mysql_connect ("$servername","$dbuser","$dbpassword");
```

Example

```
$connect = mysql_connect("localhost", "root", "mce") or die("Please, check your server
connection.");
```

where l*ocalhost* designates that the MySQL database in use is maintained on the local server, *root* is the *userid,* and *mce* is the *password*.

Note: In this book, we assume that the MySQL database server root password is mce. Please replace mce wherever it occurs in the code with the actual root password of your server.

The variable, *localhost,* will be replaced by the IP address of the MySQL server. The command, *die,* is the used to exit from the script. If the *mysql_connect()* function fails and is unable to connect to the database, the function will return *false*, display the message *Please, check your server connection,* and exit the PHP script via the *die* command. If the connection to the database is established, a *link identifier* is returned that will be stored in the variable *$connect.*

> Note: When two statements are connected with the or operator, then the second statement is executed only if the first statement returns false.

mysql_select_db()

The *mysql_select_db()* function selects the MySQL database to make it active.

Syntax

```
bool mysql_select_db ( string $database_name [, resource $link_identifier ] )
```

The function selects the active database on the server with which connection is established via a *link identifier*. If the *link identifier* is not specified, the last link opened by *mysql_connect()* is assumed. An error is displayed if a connection is not established. The function returns *true* if the database is found in the MySQL server, else returns *false*.

Examples:

```
mysql_select_db("shopping",$connect);
```

```
mysql_select_db("shopping");
```

The first example selects and makes active the *shopping* database using the connection specified in the link identifier *$connect*. The second example also selects the shopping database using the last link opened by the *mysql_connect()* function.

mysql_query()

The *mysql_query()* function sends the query to the currently active database. Rows fetched from the database table on execution of the query are stored in an array.

Syntax

```
result mysql_query ( string $query [, $link_identifier ] )
```

where *link_identifier* identifies the connection established with MySQL. If the link identifier is not specified, the last link opened by *mysql_connect()* is assumed. If no such link is found, it tries to create one by executing *mysql_connect()* with no arguments. An error is displayed if a connection is not established.

The returned result may be passed to *mysql_fetch_array()* to display each fetched row.

Example

```
$result = mysql_query($query) or die(mysql_error());
```

The *$result* returned by *mysql_query()* function is an array that contains all the rows fetched from the table that satisfies the specified query. The *$result* row array is then passed to the *mysql_fetch_array()* function for the purpose of extracting a single row to further process.

mysql_numrows()

The *mysql_numrows()* function is used to enumerate rows in a result set fetched when SQL SELECT is executed.

Syntax

```
mysql_numrows(resultset name);
```

Example

```
$num=mysql_numrows($result);
```

This example sets the value of *$num* variable equal to the number of rows stored in *$result*— the array that stores the rows retrieved from a table after an SQL Select statement is executed.

mysql_fetch_array()

The *mysql_fetch_array()* function fetches one row at a time from the row array created when a query is executed. The row can be returned either as an associative or numeric array. We can fetch one row at a time from the resultset. The function returns *false* when there are no more rows left in the resultset.

Syntax

```
row=mysql_fetch_array(data,array_type)
```

where

- *data* is the resultset containing rows of the database table created when the *mysql_query()* function is exectuted

- *array_type* (optional) specifies what kind of array to return. Its value can be:

  MYSQL_ASSOC—Returns associative array

  MYSQL_NUM—Returns numeric array

  MYSQL_BOTH— Returns both associative and numeric arrays (default)

Note: After a row is retrieved, this function automatically moves to the next row of the resultset. Each subsequent call to mysql_fetch_array() *returns the next row of the resultset.*

Example

```
$row = mysql_fetch_array($results)
```

One row is fetched from the resultset, *$results,* and stored in variable *$row.*

extract()

The *extract()* function is used to extract all the elements stored in the specified array. With this function, we can extract the data in all the fields of the specified row.

Syntax

```
extract(array name)
```

Example

```
extract($row);
```

If we assume *$row* to be a row called *customers* consisting of two fields, *userid* and *address*, then in the example, two variables, *$userid* and *$address,* are generated containing the data stored in the two respective fields of that particular row .

mysql_close()

The *mysql_close()* function closes the connection to the database server. It's a good idea to close the server connection when database operations are complete to maintain the efficiency of the web host.

Syntax

```
mysql_close()
```

4.6 Fetching Rows from Server-Side Database Tables

We've explored the functions required by our PHP scripts, so let's go ahead and use them to grab information stored in server-side database tables.

For our Book Store application, we created a *shopping* database and four tables called *books, customers, orders,* and *orders_details*. The first table, *books*, currently has a few dummy entries sorted into categories and subcategories. The last three tables are currently empty and will be filled by customer data when the user places an order.

Recall that the structure of the *books* table includes category and subcategory fields to which each book belongs (see Figure 1.1). Let's write a PHP script that fetches rows from the *books* table and displays all the available book categories. The script is shown in Listing 4.1.

Listing 4.1. PHP script for fetching and displaying book categories from the *books* table

```php
<?php
    $connect=mysql_connect("localhost","root", "mce") or die ("Please check your server
        connection");
    mysql_select_db("shopping");
    $query="Select distinct category from books";
    $results =mysql_query($query) or die (mysql_query());
    if(mysql_num_rows($results)==0)
    {
        echo '
            <ul>
                <li>No books found</li>
            </ul>';
    }
    else
    {
        echo '<ul class="rounded">';
        while ($row=mysql_fetch_array($results))
        {
```

```
        extract ($row);
        echo '<li> . $category . '</li>';
    }
    echo '</ul>';
    }
?>
```

The sequence of events is as follows:

1. A connection is made to the MSQLdatabase server.

2. Connection information is returned in the form of a *link identifier* stored in the *$connect* variable.

3. The *shopping* database is activated, so its tables can be referenced in SQLstatements.

4. An SQL statement that SELECTs distinct categories of the books from the *books* table is written and stored in a variable *$query*.

5. the SQL statement stored in *$query* is executed and rows are fetched from the *books* table. The rows containing the book categories are stored in the *$results* array.

6. We check whether the number of rows in the *$results* array is 0. If so, there was no book in the *books* table and we display the message: *No books found* on the panel.

7. If there are rows in the *$result* array, we fetch each row containing a book category, extract it to display the *$category* field via *list items*.

4.7 Inserting Rows in Server-Side Database Tables

Now let's proceed in the reverse direction to see how to store information in server-side databases. In Chapter 3, we created a form within the *Create Account* panel that prompts the user to create an account. The form has three input text fields: *Userid, ContactNo, and EmailAddress*. We also saw how the information entered in the form's input fields can be sent to a target PHP script when the Submit button is tapped. The PHP script that we created in Chapter 3 only printed a Welcome message. In this section, we'll learn how to save the information entered in the form's input fields.

Recall that when the user selects the form's Submit button, the data entered in the text fields is passed to the target PHP script. The data can be accessed from any of these three methods:

* $_POST array (if the action method used by the form is POST)

* $_GET array (if the action method used by the form is GET)

* $_REQUEST array (if the action method is either GET or POST)

The complete script is shown in Listing 4.2.

Listing 4.2. PHP script for storing customer information in a server-side database table

```
<?php
    $uid =trim($_REQUEST['Userid']);

    $contactno =trim($_REQUEST['ContactNo']);

    $emailid =trim($_REQUEST['EmailAddress']);

    $connect=mysql_connect("localhost","root", "mce") or die ("Please check your server
        connection");

    mysql_select_db("shopping");
```

```php
$query="INSERT INTO customers (userid, emailid, contact_no) VALUES('$uid',
    '$emailid','$contactno')";
$results =mysql_query($query);
echo '<p>Congratulations '  . $uid . '!. Your account is created </p>';
?>
```

The sequence of events is as follows:

- Data entered in the *Userid, ContactNo,* and *EmailAddress* input fields is accessed by using $_REQUEST array.

- The data is stored in the three variables $*uid*, $*contactno,* and $*emailid* respectively.

- We connect to the MySQL database server and select the *shopping* database to make it active.

- We write an SQL statement in the $*query* variable that will insert a row into the *customers* table. The row consists of data the user entered in the form and collected in the $*uid*, $*contactno*, and $*emailid* variables.

- Finally, the SQL statement in the $*query* variable is executed, which inserts a row in the *customers* table and displays a Congratulations message.

4.8 Querying Server-Side Database Tables

When a user selects a book category from the category list shown in Figure 1.6(b), the subcategories are fetched from server-side database tables and displayed on the screen (see Figure 1.6(c)). Let's take a look at the process of querying the tables and fetching rows that meet the given criteria.

Let's assume that the user has selected a book category, which has been passed to the PHP script. The script then fetches the appropriate subcategory and displays it on the screen. The complete script is shown in Listing 4.3.

Listing 4.3. PHP script for fetching and displaying book subcategories

```php
<?php
$category=$_POST['category'];
$connect=mysql_connect("localhost","root", "mce") or die ("Please check your
    server connection");
mysql_select_db("shopping");
$query="Select distinct subcategory from books where category ='$category'";
$results =mysql_query($query) or die (mysql_error());
if($results)
{
    echo '<ul class="rounded">';
    while ($row=mysql_fetch_array($results))
    {
        extract ($row);
        echo '<li>' . $subcategory . '</li>';
    }
    echo "</ul>";
```

```
    }
    else
    {
        echo '<ul class="rounded">';
        echo "<li> No Sub Categories found </li>";
        echo "</ul>";
    }
?>
```

The code in Listing 4.3 is exactly the same as that in Listing 4.1, with two small differences. First, in Listing 4.1, the SQL statement fetches categories from the *books* table. Listing 4.3 fetches subcategories from the *books* table. Second, only the *subcategory* column matching the given category is fetched from the *books* table and displayed on the screen as list items.

Using the same technique shown above, we can also update and delete rows in server-side database tables.

4.9 Session Management

No web application is complete without session management—a concept that helps track visitors to our web application.

HTTP is a stateless protocol, so the interaction between browsers and web servers is stateless. Each HTTP request a browser sends to a web server is independent of any other request. The stateless nature of HTTP allows users to browse the web by following hypertext links and visiting pages in any order. HTTP also allows applications to distribute or even replicate content across multiple servers to balance the load generated by a high number of requests.

Sometimes though, we need to remember specific data between web pages, for example, cart items, the user's area of interest, login information, and so on. Otherwise, the user will be asked to enter the same information each time a web page is revisited.

4.9.1 What Is a Session And How Does It Start?

A session is a combination of a server-side file containing all the data we wish to store, and a client-side cookie containing a reference to the server data. The file and the client-side cookie are created using the PHP function *session_start()*.

session_start()

This function initializes session data. It creates a session or resumes the current one based on the current session id. It has no parameters, but informs the server that sessions will be used.

Syntax

```
session_start()
```

When we call *session_start()*, PHP checks to see whether the user sent a session cookie. if the answer is *yes*, PHP loads the session data. Otherwise, PHP creates a new session file on the server, and sends an ID back to the user to associate the user with the new file. Because each user has unique data locked away in their personal session file, we need to call *session_start()* before trying to read session variables. We will soon see the role of session variables in passing data from one page to another.

session_id()

This function is used to get or set the session id for the current session.

Syntax

```
string session_id ([ string $id ] )
```

session_id() returns the session id for the current session, or the empty string ("") if the session does not exist. If an id is specified, it will replace the current session id.

To make the concept clearer, take a look at the following code:

```php
<?php
    session_start();
    $sessid=session_id();
    $_SESSION['userid'] = $_POST['userid'];
    $_SESSION['password'] = $_POST['password'];
?>
```

Here we see that first the session is started and the session id is stored in the *$sessid* variable . The session id will be used for tracking the user on different web pages of the same application. After that, we set the array elements of *$_SESSION array*. Data entered in the form's *userid* and *password* input fields on the web page calling the PHP script is collected using *$_POST array*. The data is then sent to the $_SESSION array and stored in the *$_SESSION['userid']* and *$_SESSION['password']* elements.

On any other web page of the same application, we can fetch the information stored in the $_SESSION array. The following code further illustrates the concept:

```php
<?php
    session_start();
    $uid=$_SESSION['userid'];
    $pwd=$_SESSION['password'];
?>
```

Here, after starting the session, the data stored in the *$_SESSION['userid']* and *$_SESSION['password']* elements of the *$_SESSION* array is retrieved and stored in the *$uid* and variables. These variables may be used in the current web page. This way, we can pass data across the web pages.

4.9.2 localStorage and sessionStorage

HTML5 provides two methods of storing and sharing information on the client side. HTML5 provides two objects, *localStorage* and *sessionStorage,* for passing data among web pages. Both methods stores the data in the form of key/value pairs.

localStorage

localStorage is part of the *Web Storage* specification and provides a simple JavaScript API for storing data in terms of key/value pairs in the browser.

sessionStorage

sessionStorage exists as a property of the window object and can be used for storing data in key/value pairs. The data will persist as long as that window or tab is open. The data will still be available if we navigate away from the page where we stored the data and return to the page later.

Here are some practical examples of *localStorage*.

Setting Values

To assign values to a *localStorage* key, we can use the following code snippet:

```
localStorage.userid = $('#userid').val();

localStorage.password = $('#password').val();
```

These two lines extract data entered into the web page elements *userid* and *password,* and store the extracted data in the *userid* and *password localStorage* keys.

To assign values stored in JavaScript variables to the *localStorage* keys, we can use the following:

```
localStorage.userid =usr;

localStorage.password = pswd;
```

The values in the *usr* and *pswd* JavaScript variables are stored in the *localStorage userid* and *password* keys.

Getting Values

If we want to retrieve the values stored in *localStorage* keys on any web page, we can use the following:

```
var userid = localStorage.getItem('userid');

var password = localStorage.getItem('password');
```

or

```
var userid = localStorage.userid;

var password = localStorage.password;
```

In both snippets, data stored in the *userid* and *password localStorage* keys are retrieved and stored in the *userid* and *password* JavaScript variables.

To assign the data stored in the *localStorage* keys to web page elements with specific ids, we use following:

```
$('#userid').val(localStorage.userid);

$('#password').val(localStorage.password);
```

Data stored in the *userid* and *password localStorage* keys are assigned to the web page's *userid* and *password* elements .

Deleting Values

To delete a specific key/value pair from storage, we can use either of the two following snippets:

```
localStorage.removeItem('userid');
```

or

```
delete localStorage.userid;
```

Both examples delete the *userid* key value pair from *localStorage*.

To delete all key/value pairs, we use this snippet:

```
localStorage.clear();
```

All the above examples can be used with *sessionStorage* as well. Just replace *localStorage* with *sessionStorage*. *localStorage* and *sessionStorage* differ only in terms of persistence and scope. We can have the same keys in the *localStorage* and *sessionStorage* as well, because both sets are stored separately.

4.10 Summary

In this chapter, we saw how client-side and server-side database tables are created and how information is inserted, fetched and updated.

We also examined the uses of session management in tracking a visitor and how the data can be passed among other web pages. Finally, we saw how *localStorage* and *sessionStorage* can be used for storing information on the client and used across web pages

In the next chapter we'll learn about AJAX and how it helps in asynchronous transmission from the web server, making our web application highly responsive.

5

AJAX

In the previous chapter, we saw how the information is maintained in client-side and server-side database tables. It's now time to address the concept of AJAX and the benefit of implementing it in a web application. AJAX is a technique that makes a web page highly responsive, because it allows asynchronous requests to be made back to the server without the need to reload pages. In this chapter, we are going to learn about:

- AJAX technology

- Limitations of traditional web applications

- Components of AJAX

- How to implement AJAX in our sample *Book Store* web application

5.1 Introduction to AJAX

Ajax stands for **Asynchronous JavaScript and XML**. In a traditional web application, when we want to access data from server-side database, an HTTP request from the client is made by either the GET or POST method. After receiving the data from the server, the web page needs to be reloaded to show the fetched data. With AJAX, we can request and receive server data in the background and display it without reloading. AJAX allows JavaScript to communicate directly with the server via a JavaScript XMLHttpRequest.

Because of the round-tripping, a traditional web page may take longer to properly display results. User-entered information is sent from the client to the web server, which processes the data and returns it to the client. This process takes place regardless of the number of changes made by the user. There's no facility to refresh a web page fragment, so the entire page must be refreshed, and users don't like to wait.

There are three steps of communication performed in traditional web application mode (see Figure 5.1):

- The client makes an HTTP request to the web server.

- The web server searches for the desired data from the database and

- The fetched data from the database is returned to the client by the web server in the form of a response. The entire page is then reloaded with the information sent by the web server.

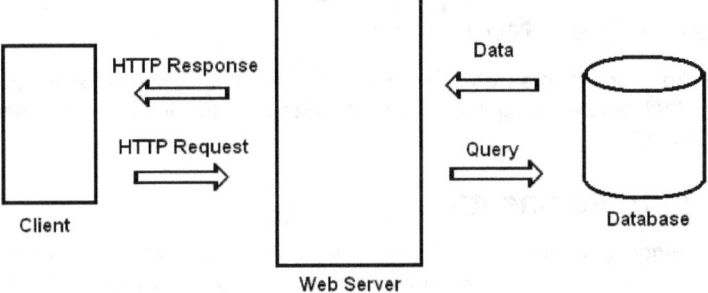

Figure 5.1. The Traditional Web Application Model

Limitations of Traditional Web Applications

Here are three disadvantages to a conventional web application:

- All the form data is sent to the web server, even if small changes were made.

- Transfer of large data blocks during postback results in network congestion.

- Until and unless the user clicks a button or other control that posts data back to the server, no result will be displayed.

With AJAX, we don't face any of these limitations. Not only can we get a server response without refreshing the entire page, but sections may be refreshed independently. Let's take a look at AJAX components before we see how to implement it.

5.2 AJAX Components

AJAX consists of following components:

- XmlHttpRequest

- JavaScript

- DOM (Document Object Model)

- CSS (Cascading Style Sheets)

XmlHttpRequest

This object is used to talk to the server asynchronously—allowing the browser to talk to the server without requiring a *postback* of the entire web page. How this capability is provided varies from browser to browser. For example, in Internet Explorer, this capability is provided by the *MSXML ActiveX* component. With Mozilla Firefox and other web browsers, this capability is provided by the object called *XmlHttpRequest*.

JavaScript

Ajax applications use JavaScript code for the following reasons:

- It's an interpreted scripting language

- The command syntax is easy to learn

- It can automate multiple tasks, such as user id and email validation.

Note: JavaScript is the client-side scripting language supported by all major browsers.

DOM (Document Object Model)

DOM provides a tree-like structure to a web page and makes the web page elements appear as a set of programmable objects that can be manipulated using JavaScript code. DOM allows dynamic updating of an individual web page section.

CSS (Cascading Style Sheets)

CSS is a centralized method of defining the styles we want to apply to different elements on a web page. It makes web applications appear more consistent and attractive. CSS styles are implemented by defining style rules in a separate document, which are then referred to by the web page when styles need to be applied.

AJAX model Communication

In AJAX, the client makes use of an *XMLHTTP Request* object to communicate with the web server for fetching data without postback. The client may use:

- JavaScript to automate validation or navigation tasks
- CSS for applying styles uniformly
- DOM to update a part of the web page

The web server searches for the desired data in the database and the fetched data, which may only be a part of the entire page, is returned to the client to quickly update that section.

In summary, then, Ajax applications make use of an *XmlHttpRequest* object to talk to the server asynchronously and retrieve only the requested data. And, on the client-side, JavaScript processes the server's response and modifies the document contents through its DOM to specify that the action is complete. CSS provides a consistent look and feel to the web application.

5.3 Implementing AJAX

We will be implementing AJAX in our Book Store application wherever we need to insert or fetch data from the server-side database tables, thus improving performance.

Let's say we want to display detailed book information whose ISBN number has been entered by the user. This data is already stored in the *books* server-side database table.

We'll need three things to fetch the data:

- A function to invoke the AJAX method
- A server-side PHP script
- A div element to display the response

5.3.1 Invoking the AJAX Method

To implement AJAX, we need to invoke the ajax() method, and we will do so by creating a function called *showdetails*. The user-entered ISBN number is passed to *showdetails* as a parameter. The code of *showdetails* is shown in Listing 5.1.

Listing 5.1. Invoking the ajax() Method

```
function showdetails(isbn)
{
    $.ajax({
        type:"POST",
        url:"getdetails.php",
        data: 'isbn='+isbn,
        success:function(html){
            $('#bookdetails').append(html);
            jQT.goTo('#bookdetailsdisplay', 'slide');
        }
```

```
    });

    return false;

}
```

The ajax() method loads a remote page via an HTTP request, and also creates and returns the *XMLHttpRequest object* required to talk to the server asynchronously. The *ajax()* method takes one argument, an object of key/value pairs, which are used to initialize and handle the request. The ajax() method syntax is :

```
.ajax( object of key/value pairs )
```

A brief description of key/value pairs frequently used in the ajax() method are shown in Table 5.1.

Table 5.1. Key Value Pairs Used in the ajax() Method

Key	Description
type	A string that defines the HTTP method we used for the request—GET or POST. The default is GET.
url	A string containing the URL of the server script file to which we want to send the request
data	A map or string we want sent to the server script along with the request. The script will then process the sent data.
	To create the data map, we retrieve the user-entered information and assign it to variables in the following format:
	`data:'variable1=value1&variable2=value2.....';`
	Here's a snippet of how the data map is used:
	`var cat=$('.category').val();`
	`var subcat=$('.subcategory').val();`
	`data: 'category='+cat+'&subcategory='+subcat,`
	We assume the Form on our web page contains two input fields, *category* and *subcategory*. The information entered into the two fields is retrieved and stored in the variables *cat* and *subcat* respectively. Thereafter, the *data* map is created by using these two variables
	We can also assign all the `variable=value` pairs to a variable and then use it to make our *data* map, as shown below:
	`var data='category='+cat+'&subcategory='+subcat;`
	`data: data,`
	If this *data* variable exists, it's is sent to the server to be assigned to the specified script file, which performs its function and generates a response. We can skip the *data* variable if we don't want any information passed to the script.
success	A callback function executed if the request sent to the script succeeds. The returned response from the server script is assigned to this callback function's parameter.

The steps involved in Listing 5.1 are as follows:

1. The code in the *showdetails* function invokes the request via the *ajax()* method.

2. In *ajax()*, we specify that the request method we will use is POST and the name of the server-side script file that will be executed is *getdetails.php*.

3. We create a *data* string to be passed to the PHP script, consisting of an *isbn* variable set to the value of the *isbn* passed to the function.

4. The *data* string, in turn, passes the *isbn* number to the script file, *getdetails.php.*

5. This script fetches the corresponding ISBN information from the *books* table and generates the appropriate output. The *success* callback function will execute if the request passed to the script succeeds. The output of the script will be assigned to the callback function's *html* parameter.

6. The contents of the *html* parameter is then appended to the *bookdetails* div element to display the result. We assume that there exists a *bookdetails* div nested inside the *bookdetailsdisplay* div.

7. The statement, *jQT.goTo('#bookdetailsdisplay', 'slide');* navigates to the *bookdetailsdisplay* panel with a slide animation to display the response generated by the *getdetails.php* script.

8. Finally, we return a value of *false* to the *showdetails* function to suppress the default browser click behavior. We explain this code in more detail below.

Let's assume that *showdetails* is invoked by a click event on the button created here:

```
<a class="whiteButton" href="#" onclick="showdetails('11111-1111-1111');"> Show
    Details</a>
```

A button, called *Show Details*, is created by assigning a *whiteButton* to the hyperlink. When clicked, this button invokes the *showdetails* function with a parameter, '11111-1111-1111,' that has been passed to it. We want to suppress the default browser click behavior, so, we return *false* from the *showdetails* function.

5.3.2 The Server side PHP Script

The server-side PHP script is the core of the fetch process. The script file, *getdetails.php,* reads the *isbn* number passed to it via *data* and will use this value to fetch the detailed book information from the *books* table. The script will the generate output for display in the panel. The code of *getdetails.php* is shown in listing 5.2.

Listing 5.2. getdetails.php

```php
<?php
    $isbn =trim($_REQUEST['isbn']);
    $connect=mysql_connect("localhost","root", "mce") or die ("Please check your server
        connection");
    mysql_select_db("shopping");
    $query="Select isbn, title, author1, author2, author3, publisher,
        publish_date_edition, price, image, description from books where isbn ='$isbn'";
    $results =mysql_query($query) or die (mysql_query());
    while ($row=mysql_fetch_array($results))
    {
        extract ($row);
        echo '<fieldset style="background-color:white; color:black;">';
        echo '<img src=' . $image .'>';
        echo '<h3>' . $title . ' by </h3>';
        echo '<h4>' . $author1 . '</h4>';
        if($author2 !='NULL')
```

```php
        echo '<h4>' . $author2 . '</h4>';
    if($author3 !='NULL')
        echo '<h4>' . $author3 . '</h4>';
    echo '<label>Publisher :</label><h4>' . $publisher . '</h4>';
    echo '<h4>' . $publish_date_edition . '</h4>';
    echo '<label>Price: </label>';
    echo '<em>' . $price . '</em><br/>';
    echo '<label>Book Details :</label><h4>' . $description . '</h4>';
    echo '</fieldset>';
    }
?>
```

We can see that the above code of the script does the following:

1. Retrieves the value of *isbn* from the *$_POST* array and stores it in the variable *$isbn*.
2. Establishes a connection to the MySQL server using *root* as the user and *mce* as the password.
3. Selects the shopping database to make it active.
4. Writes a query to fetch the information for the book having the supplied ISBN number.
5. Executes the query and receives the result in a *$result array*.
6. Retrieves a row from *$result* and stores it into a *$row* variable.
7. Extracts the *$row* containing the retrieved row fields.
8. Displays the information in the $row's fields in a fieldset.

Remember that the detailed information displayed by this PHP script will be returned to the *success* callback function's *html* parameter (see Listing 5.1). The detailed book information assigned to the *html* parameter is then appended to the *bookdetails* div element for display. Let's take look at the code of the div elements.

5.3.3 Displaying Responses with the div Element

bookdetails div is nested inside the parent *bookdetailsdisplay* div element. The *bookdetailsdisplay* div element is the main panel. This element contains two divs: *bookdetails,* used to display the response generated by the PHP script, and *toolbar,* which displays the toolbar in the panel. The toolbar has one button, *Back*, for returning to the previous panel, and a heading *Book Details,* that's displayed as panel title. The *bookdetailsdisplay* div code is shown in Listing 5.3.

Listing 5.3. Displaying the PHP Script Response

```html
<div id="bookdetailsdisplay">
    <div class="toolbar">
        <a class="back"  href="#">Back</a>
        <h1>Book Details</h1>
    </div>
```

```
<div id="bookdetails">
</div>
</div>
```

5.4 Summary

In this chapter, we discussed the limitations of traditional web applications and the components of AJAX. We saw how AJAX can be used to access information from server-side database tables. In the next chapter, we'll further develop the entire Book Store application, one step at a time.

6

Assembling the Store, Part 1: Displaying Store Contents

In previous chapters, we've looked at a variety of bits and pieces that will be used to assemble our sample bookstore application. The next three chapters tie together all this information into a complete program. We're going to develop the bookstore in three steps:

* **Displaying the store contents**. In this chapter, we'll learn how to create the *Home* panel, display book categories, subcategories, and detailed information about each book, including discount offers and new arrivals.

* **Maintaining the Cart**. In Chapter 7, we'll write the code for adding, updating, and deleting cart items.

* **Placing Orders.** In Chapter 8, we'll write the code for creating accounts, signing in, placing orders, and obtaining shipping information from the user.

This chapter is focused on displaying the contents of the store, so we'll be creating five panels:

* The *Home* panel, which lists the application's main menu items.

* The *Categories* panel, which displays book categories

* The *Subcategories* panel, which displays book subcategories.

* The *Select Books* panel, which lets users browse a list of books and add them to the shopping cart.

* The *Book Details* panel, which displayed displays detailed book information and allows users to add them to the shopping cart.

6.1 Creating the *Home* Panel

The Home panel, shown in Figure 6.1, is the heart of the application. This page displays the main options available to the user—Books, New Arrivals, Contact Us, and so on.

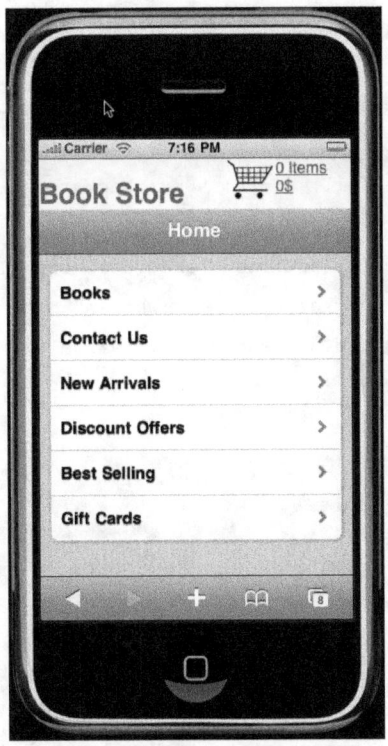

Figure 6.1. The *Home* panel screen

The Home Panel is arranged thusly:

- The *home* div displays a header, *Book Store,* above the toolbar in 29-pixel bold, blue *Helvetica* font.

- After the header, a cart icon is displayed followed by two div elements, *totqty* and *totprice*. These display the quantity of each book in the cart and their total price. The values of *totqty* and *totprice* will be assigned by the *readcart()* function. When either of the div element is selected, we will jump to the *showcart* div using a *slide* animation. These div elements also invoke the *dispcart()* function, used to display the cart contents. We will be creating the *showcart div* and *dispcart()*function in Chapter 7.

- Underneath the cart icon is the *toolbar* div, whose main job is to display the panel title and navigation buttons. We don't want any navigation buttons in the *Home* toolbar panel, so the div contains only the *Heading 1* element, which displays the title, *Home*.

- Below the toolbar is an unordered list of six items: *Books, Contact Us, New Arrivals, Discount Offers, Best Selling,* and *Gift Cards*. Each list item contains an *href* hyperlink that points to the appropriate panel. We'll cover the *Books* list in the next few sections. The code for the remaining list items (*contactus, newarrivals, discountoffers, bestselling, and giftcards),* is discussed in the *Coding the Remaining Home Panel List Items* section of this chapter.

To display the Home panel, we need to create the following:

- A div element called *home,* used to display the Home Panel itself (see Listing 6.1). The *home* div is the first panel of our web application and will be displayed when the web application starts.

- A JavaScript function called *readcart(),* which displays cart contents and pricing (see Listing 6.2).

Listing 6.1. *home* div

```html
<div id="home">
    <h1 style="display: table-cell; width: 190px;background-color:white;
        color: #blue; font: bold 28px Helvetica;"> Book Store </h1>
    <div style="display: table-cell; width: 50px; background-color:white;">
        <img  src="cartfigure.tiff"> </div>
    <a style="display: table-cell; position: absolute; top: 1px; height:50px;
        right:0px; left: 244px; background-color:white;" href="#"
        onclick="dispcart(); jQT.goTo('#showcart', 'slide');">
        <div class="totqty"></div>
        <div class="totprice"></div>
    </a>
    <div class="toolbar">
        <h1>Home</h1>
    </div>
    <ul class="rounded">
        <li class="arrow"><a href="#" onclick="showcategories();">Books</a></li>
        <li class="arrow"><a href="#contactus">Contact Us</a></li>
        <li class="arrow"><a href="#newarrivals">New Arrivals</a></li>
        <li class="arrow"><a href="#discountoffers">Discount Offers</a></li>
        <li class="arrow"><a href="#bestselling">Best Selling</a></li>
        <li class="arrow"><a href="#giftcards">Gift Cards</a></li>
    </ul>
</div>
```

Listing 6.2. *readcart()* function

```javascript
function readcart()
{
    var total=0;
    var subtot=0;
    var qtycount=0;
    var sid="<?php  echo $sessid; ?>";
    var datab;
    var shortName = 'tmpCart';
    var version = '1.0';
    var displayName = 'tmpCart';
    var maxSize = 200000;
    datab = openDatabase(shortName, version, displayName, maxSize);
```

```
$('.totqty').children().remove();
$('.totprice').children().remove();
datab.transaction(
    function(transaction) {
        transaction.executeSql(
            'SELECT cart_sess, cart_isbn,  cart_item_name, cart_qty, cart_price
                FROM shopcart where cart_sess=?;',[sid],
            function (transaction, result) {
                if (result.rows.length <=0)
                {
                    $('.totqty').append('<p>0 Items</p>');
                    $('.totprice').append('<p>0$</p>');
                }
                else
                {
                    for (var i=0; i < result.rows.length; i++)
                    {
                        subtot=0;
                        var row = result.rows.item(i);
                        qtycount=qtycount+parseInt(row.cart_qty);
                        subtot=row.cart_qty*row.cart_price;
                        total=total+subtot;
                    }
                    $('.totqty').append('<p>' + qtycount +' Items</p>');
                    $('.totprice').append('<p>' + total.toFixed(2) +' $</p>');
                }
            },
            displayerrormessage
        );
    }
);
}
```

The sequence of events in Listing 6.2 is as follows:

- Three variables, *total, subtot,* and *qtycount,* are initialized to 0.

- The *subtot* variable totals the price of all the copies of a particular book by multiplying quantity (copies) by the price of a single copy.

- The *qtycount* variable counts the total copies of each book.

- The *total* variable stores the sum of all the *subtot* variables—the total cost of books in the cart.

- A connection to the client-side database, *tmpCart,* is established. If the local database doesn't already exist by the specified name, a new one is created.

- A session id of the user is created by following code which will be used in recognizing his cart

```php
<?php

    session_start();

    $sessid=session_id();

?>
```

- The session id code shown above and in Listing 6.1 and Listing 6.2 is the part of the *Index.php* script, which is the first file that executes when the web application runs (the complete source code is shown in Appendix C). This script also links the rest of the application's pages together.

- The session id in the *$sessid* variable is assigned to the JavaScript variable *sid*.

- Any *quantity* or *amount* displayed earlier in the *totqty* and *totprice* div elements are removed. A search is made in the *shopcart* table of the *tmpCart* database to see if there's anything in the cart. If the number of rows in the result is 0, nothing is in the cart and the *0 Items 0$* text is displayed in the header besides the cart icon.

- If the *shopcart* table is not empty, then each of row is fetched into a *row* variable and the value in the *cart_qty* fields is added into the variable *qtycount*.

- The subtotal is computed by multiplying the *cart_qty* and *cart_price* fields and the result is stored in *subtot* variable.

- The subtotal of each book is added into the *total* variable .

- The values in the *totqty* variable and *totprice* variables are displayed in the panel next to the cart icon via the *totqty* and *totprice* div elements.

6.2 Creating the *Categories* Panel

When a user taps the list item *Books*, we want to display a list of categories. Category information has to be fetched from the *books* table in the server-side database, *shopping*, and presented to the user as another list, as shown in Figure 6.2.

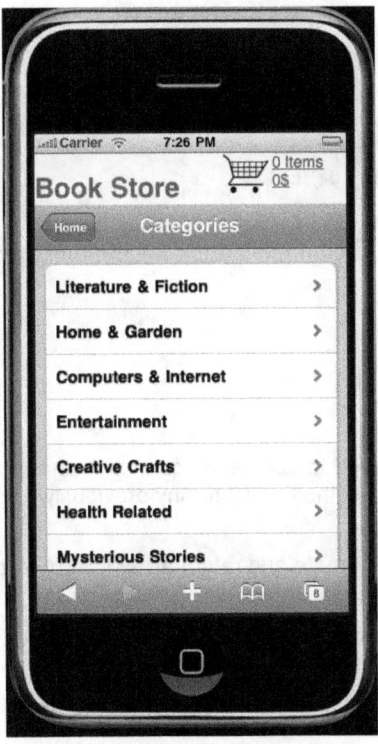

Figure 6.2. *Categories* panel showing book categories

To display different categories of books, we need the following:

- The JavaScript function *readcart()*, which was created in the *Home* panel code
- The JavaScript function *showcategories()*, which manages the whole process
- The PHP script *getcategories.php*, which fetches categories from the *books* table
- A *bookscategories* div element, which displays the categories on the screen

Let's first create the JavaScript function *showcategories()*, which will perform the following tasks:

- Calls the server-side *getcategories.php* PHP script, which fetches the categories from the server-side *books* database table.
- Displays the book categories fetched by the PHP script via the *bookscategories* div.
- Navigates to the *bookscategories* div to display the categories.

The code of the *showcategories()* function is shown in Listing 6.3.

Listing 6.3. *showcategories()* function

```
function showcategories(){
    readcart();
    $('#categories').children().remove();
```

```
$.ajax({

    type:"POST",

    url:"getcategories.php",

    success:function(html){

        $('#categories').append(html);

        jQT.goTo('#bookscategories', 'slide');

    }

});

return false;

}
```

The sequence of events is as follows:

1. The *readcart()* function is called to update the quantities and price of books in the cart (if any) displayed in the header next to the cart icon. Any previously displayed book category is removed from the *categories* div.

2. The *ajax()* method is used to create and return an *XMLHttpRequest* object, which is required to talk to the server asynchronously.

3. Through the *ajax()* method, we specify the Post request method and the *getcategories.php* as the name of the server-side script we want executed.

4. The script file, *getcategories.php,* will fetch the available book categories from the server-side *books* database table.

5. When the request sent to *getcategories.php* succeeds, the *success* call back function executes and the script response will be received in its *html* parameter.

6. The contents of the *html* parameter is appended to the *div* element *categories*, which is nested inside the *bookscategories* div element, as shown in Listing 6.5.

7. The statement, *jQT.goTo('#bookscategories', 'slide');* send us to the *bookscategories* panel with a *slide* animation, displaying the response generated by *getcategories.php*.

8. We return *false* in the function to suppress the default browser click behavior.

The server-side PHP script, *getcategories.php*, is the means by which category information is fetched from the *books* table. The code is shown in Listing 6.4.

Listing 6.4. *getcategories.php* script

```php
<?php
    $connect=mysql_connect("localhost","root", "mce") or die ("Please check your server
        connection");
    mysql_select_db("shopping");
    $query="Select distinct category from books";
    $results =mysql_query($query) or die (mysql_query());
    if(mysql_num_rows($results)==0)
    {
```

```
        echo '
            <ul>
                <li>No books found</li>
            </ul>';
    }
    else
    {
        echo '<ul class="rounded">';
        while ($row=mysql_fetch_array($results))
        {
            extract ($row);
            echo '
                <li class="arrow"><a href="#" onclick="javascript:selectedcat(\'' .
                    urlencode($category) .'\');"> ' . $category . '</a></li>';
        }
        echo '</ul>';
    }
?>
```

The sequence of events in Listing 6.4 is as follows:

1. The connection to the server-side *shopping* database is established and a query is executed to display the unique book categories in the *books* tables.

2. The list of categories fetched from the *books* table is received in the *results* array. If the $results array has is empty, it means that there were no corresponding rows in the *books* table. In this case, we generate a *No books found* response and send it back to the *showcategories()* function.

3. If the *results* array is not empty, one row at a time is retrieved and assigned to *$row* variable.

4. The contents of the *$row* variable is extracted to display the list of categories stored in the *category* column. The actual display is accomplished by sending the response to the *showcategories()* function, which in turn appends the response to the *categories* div.

5. When the user selects a category from the list, the associated hyperlink of the selected list item invokes the *selectedcat()* function, and passes the *category* parameter to it. We will learn in the next section that *selectedcat()* function is used to display subcategories of the selected category.

6. Some categories, for example *Literature & Fiction*, contain spaces. Unfortunately, spaces act as delimiters, and text strings being passed as parameters may not include them. Spaces are handled with a function called *urlencode()*, which wraps the complete category name, including spaces, and passes the category to the *selectedcat()* function.

The categories fetched the *getcategories.php* script is displayed via the *bookscategories* div, the code for which is shown in Listing 6.5.

Listing 6.5. *bookscategories* div

```
<div id="bookscategories">

    <h1 style="display: table-cell; width: 190px;background-color:white;
        color: #blue; font: bold 28px Helvetica;">Book Store </h1>

    <div style="display: table-cell; width: 50px; background-color:white;">

        <img  src="cartfigure.tiff">

    </div>

    <a style="display: table-cell; position: absolute; top: 1px; height:50px;

        right:0px; left: 244px; background-color:white;" href="#"

        onclick="dispcart(); jQT.goTo('#showcart', 'slide');">

        <div class="totqty"></div>

        <div class="totprice"></div>

    </a>

    <div class="toolbar">

        <a class="back"  href="#home">Home</a>

        <h1>Categories</h1>

    </div>

    <div id="categories">

    </div>

</div>
```

Similar to the *home* div , the *bookscategories* div also displays a *Book Store* header above the toolbar, as well as a cart icon and the two div elements, *totqty* and *totprice,* showing the total quantity and cost of the cart contents.

Below the header is a *toolbar* div with a *Back* button to return to the *Home* panel. The toolbar contains a Heading 1 element set to display the title *Categories* on the panel. Below the toolbar is *categories* div element, containing the list of categories generated by *getcategories.php* and displayed by the *showcategories()* function.

6.3 Creating the *Subcategories* Panel

The next panel, *Subcategories,* displays book subcategories in a selected category. Again, information about the different subcategories has to be fetched from the *books* table in the server-side *shopping* database and is displayed in a list, as shown in Figure 6.3.

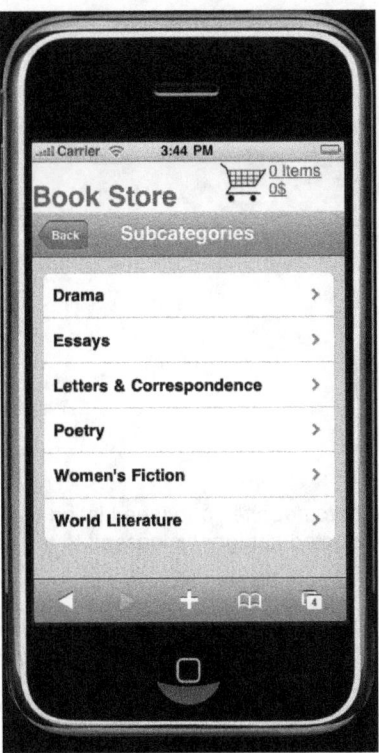

Figure 6.3. *Subcategories* panel showing book subcategories

To display different subcategories of books, we need the following:

- A JavaScript function, *readcart()*, which we already created in the *Home* panel code.
- A JavaScript function, *selectedcat()*, which manages the entire process.
- A PHP script, *getsubcat.php* used to fetch subcategories from *books* table.
- A *subcategories* div element, used to display the subcategories on the screen.

Let's first create the JavaScript function, *selectedcat()*, which will do the following:

- Call the server-side PHP script, *getsubcat.php,* to fetch subcategories from the server-side *books* database table.
- Display the subcategories fetched by the PHP script via the *subcategories* div.
- Navigate to the *subcategories* div to make the subcategories visible.

The *selectedcat()* function code is shown in Listing 6.6.

Listing 6.6. selectedcat() function

```
function selectedcat(categ){
    $('#subcat').children().remove();
    $.ajax({
```

```
        type:"POST",
        url:"getsubcat.php",
        data: 'category='+categ,
        success:function(html){
            $('#subcat').append(html);
            jQT.goTo('#subcategories', 'slide');
        }
    });
    return false;
}
```

The sequence of events is as follows:

1. If a category returned by the *getcategories.php* (see Listing 6.4) is selected, the *selectedcat()* function is invoked and the selected category is passed to the function as the *categ* parameter.

2. Any displayed subcategories of the selected category is removed from the *subcat* div, to prevent confusion between prior and current subcategories.

3. We call the *ajax()* method to create and return the *XMLHttpRequest* object—essential to implementing AJAX.

4. In the *ajax()* method, we declare the POST request method and *getsubcat.php* as the server-side script file name to which the request will be sent.

5. We pass the category to the script file via the *data* key.

6. *getsubcat.php* will fetch subcategories belonging to the category from the server-side *books* database table.

7. When the request sent to *getsubcat.php* succeeds, the *success* callback function executes and the script response is be received in its *html* parameter.

8. The contents of *html* will be appended to the *subcat* div element.

9. Because the *subcat* div element is nested inside the *subcategories* div element, we jump to the *subcategories* div using *slide* animation.

10. And finally, to suppress the default browser click event behavior, the function returns *false*.

The server-side PHP script, *getsubcat.php*, fetches the subcategories list of the selected category from the *books* table, as shown in Listing 6.7.

Listing 6.7. *getsubcat.php* script

```php
<?php
    $category=$_POST['category'];
    $connect=mysql_connect("localhost","root", "mce") or die ("Please check your server
        connection");
    mysql_select_db("shopping");
```

```php
$query="Select distinct subcategory from books where category ='$category'";
$results =mysql_query($query) or die (mysql_error());
if($results)
{
    echo '<ul class="rounded">';
    while ($row=mysql_fetch_array($results))
    {
        extract ($row);
        echo '<li class="arrow">';
        echo "<a href=\"#\" onclick=\"javascript:selectedsubcat('" .
            urlencode($category). "','". urlencode($subcategory) ."');\">" .
            $subcategory . "</a></li>";
    }
    echo "</ul>";
}
else
{
    echo '<ul class="rounded">';
    echo "<li> No Subcategories found </li>";
    echo "</ul>";

}
?>
```

The sequence of events is as follows:

1. The parameter sent to the *getsubcat.php* script via *data* is retrieved from the *$_POST* array and stored in the *$category* variable .

2. The connection to the *shopping* database is established and a query is executed to fetch from the *books* table the subcategories that matches the category specified in the *$category* variable.

3. The list of subcategories is fetched from the *books* table and stored in the *results* array.

4. If the *$results* array is empty, that is, there are no subcategories matching the query, a *No Subcategories found* response is generated and returned to the *selectedcat()* function for display.

5. If the *results* array is not empty, each row is retrieved and assigned to *$row* variable.

6. The row retrieved in *$row* variable is extracted, displaying the subcategories stored in the *subcategory* column as a list.

7. The response of the script is sent to the *selectedcat()* function, which appends the response to the *subcat* div.

8. To list the books in the subcategory on the display, the selected category and subcategory are passed to the *selectedsubcat()* function as parameters. Both parameters are encoded via *urlencode()* to insure that spaces pass through correctly. In the next section, we will see how

selectedsubcat() is used to display the list of books belonging to the selected category and subcategory.

The subcategories fetched from the *books* table are displayed by appending them to the *subcategories* div, shown in Listing 6.8.

Listing 6.8. *subcategories* div

```
<div id="subcategories">
    <h1 style="display: table-cell; width: 190px;background-color:white;
        color: #blue; font: bold 28px Helvetica;    "> Book Store </h1>
    <div style="display: table-cell; width: 50px; background-color:white;">
        <img  src="cartfigure.tiff"> </div>
    <a style="display: table-cell; position: absolute; top: 1px; height:50px;
        right:0px; left: 244px; background-color:white;" href="#"
        onclick="dispcart(); jQT.goTo('#showcart', 'slide');">
        <div class="totqty"></div>
        <div class="totprice"></div>
    </a>
    <div class="toolbar">
        <a class="back"  href="#">Back</a>
        <h1>Subcategories</h1>
    </div>
    <div id="subcat">
    </div>
</div>
```

Just like the *bookscategories* div, the *subcategories* div displays a *Book Store* header above the toolbar, along with a cart icon and the *totqty* and *totprice* div elements. Below the header is a toolbar with a *Back* button that returns the user to the *Categories* panel. The toolbar also contains a *Subcategories* Heading 1 element. Below the toolbar is a *subcat* div element that displays the subcategories generated by *getsubcat.php* and displayed via the *selectedcat()* function.

6.4 Creating the *Select Books* Panel

The next panel displays books in the selected subcategory and provides buttons for reading detailed information or dropping the book into the shopping cart.

The list of the books must be fetched from the *books* table in the server-side *shopping* database, as shown in Figure 6.4(a). There can be multiple books in the subcategory, and these are displayed sequentially. The user can scroll down to see additional entries, as shown in Figure 6.4(b).

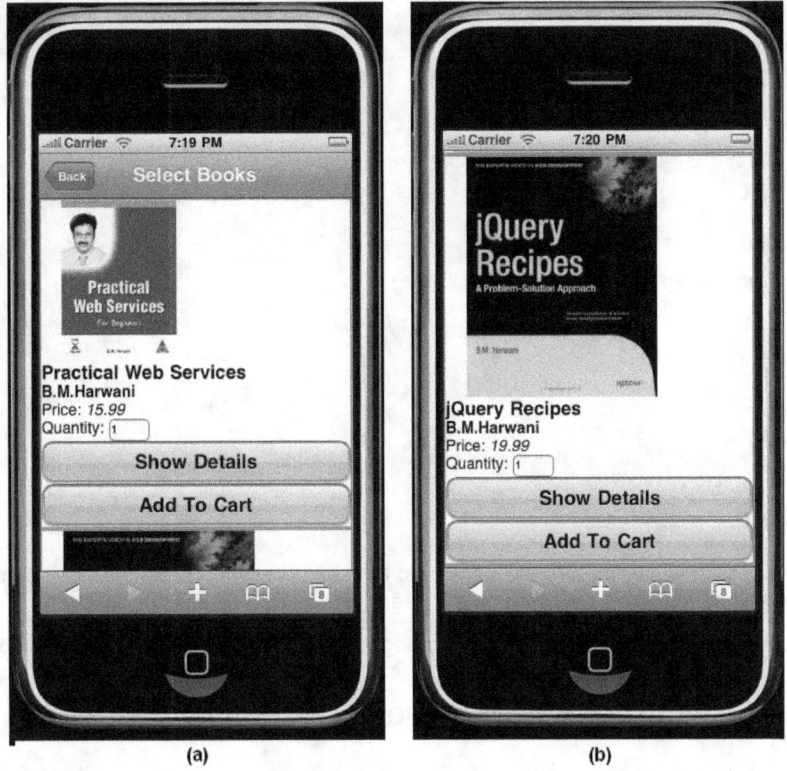

(a)   (b)

Figure 6.4. (a) List of books displayed in a selected subcategory (b) Scrolling the screen to view more books

To display all the books belonging to the selected subcategory, we need the following:

- A JavaScript function, *selectedsubcat()*, that manages the entire process

- *A getbooks.php* PHP script that fetches books in the selected subcategory from the *books* table

- A *booksdisplay* div element that displays the books on the screen

Let's first create the *selectedsubcat()* JavaScript function, which will perform the following tasks:

- Calls the server-side *getbooks.php* PHP script, which fetches books belonging to the selected subcategory from the server-side *books* database table.

- Sends the books fetched by the PHP script to the *booksdisplay* div.

- Jumps to the *booksdisplay* div, which outputs the book list to the screen.

The *selectedsubcat()* code is shown in Listing 6.9.

Listing 6.9. *selectedsubcat()* function

```
function selectedsubcat(cat, subcat){
    $('#bookslist').children().remove();
    $.ajax({
        type:"POST",
        url:"getbooks.php",
        data: 'category='+cat+'&subcategory='+subcat,
        success:function(html){
            $('#bookslist').append(html);
            jQT.goTo('#booksdisplay', 'slide');
        }
    });
    return false;
}
```

The sequence of events is as follows:

1. The *cat* and *subcat* parameters are passed to the *selectedsubcat()* function that contains the category and subcategory selected by the user.

2. Any of the books in a previously selected category and subcategory are removed from the *bookslist* div.

3. The *ajax()* method is called, requesting the POST method and the *getbooks.php* server-side script be used.

4. The selected category and subcategory is passed to the script in the variables *category* and *subcategory* as *data*.

5. The script file, *getbooks.php,* fetches the books from the server-side *books* database table that meet the criteria supplied by *data*.

6. When the request sent to *getbooks.php* succeeds, the *success* call back function executes.

7. The script response is received in an *html* parameter that is appended to the *bookslist* div.

8. The *bookslist* div is nested inside the *booksdisplay* div, hence we are navigated to it via a slide animation and displays the list of books.

9. To suppress the browser's default click-event behavior, the function returns *false*.

The code for the server-side PHP script, *getbooks.php,* is shown in Listing 6.10.

Listing 6.10. *getbooks.php* script

```
<?php
    $cat =trim($_REQUEST['category']);
    $subcat =trim($_REQUEST['subcategory']);
```

```php
$connect=mysql_connect("localhost","root", "mce") or die ("Please check your server
    connection");
mysql_select_db("shopping");
$query="Select isbn, title, author1, author2, author3, price, image from books where
    category =\"$cat\" and subcategory = \"$subcat\"";
$results =mysql_query($query) or die (mysql_query());
if(mysql_num_rows($results)>0)
{
    while ($row=mysql_fetch_array($results))
    {
        extract ($row);
        echo '<fieldset style="background-color:white; color:black;">';
        echo '<form action="cart.php?isbn=' . $isbn . '&title=' . urlencode($title) .
            '&price=' . $price .'&action=add' . '" method="POST" class="form">';
        echo '<img src=' . $image .'>';
        echo '<h3>' . $title . '</h3>';
        echo '<h4>' . $author1 . '</h4>';
        echo '<label>Price: </label>';
        echo '<em>' . $price . '</em><br/>';
        echo '<label>Quantity:  </label><input type="text" name="quantity" value="1"
            style="height:22px;" size="6"/>';
        echo '<a class="whiteButton" href="#" onclick="showdetails(\'' . $isbn .
            '\');"> Show Details</a>';
        echo '<a class="submit whiteButton" href="#"> Add To Cart</a>';
        echo '</form>';
        echo '</fieldset>';
    }
}
else
{
    echo '<ul class="rounded">';
    echo "<li> No Books found in this Subcategory</li>";
    echo "</ul>";
}
?>
```

The sequence of events is as follows:

1. The category and subcategory selected by the user is passed to the script *getbooks.php* via *data*.

2. The category and subcategory are retrieved from a *$_POST* array and stored in the *$cat* and *$subcat* variables.

3. A connection to the *shopping* database is established and a query executed to fetch information from the *books* table that match the category and subcategory specified in the *$cat* and *$subcat* variables respectively.

4. The list of books belonging to the specified category and subcategory is fetched from the *books* table and stored in the *results* array.

5. If the *$results* array is empty, that is, if there are no books of the selected category and subcategory, then a *No Books found in this Subcategory* response is generated and sent back to the *selectedsubcat()* function for display.

6. If the *results* array is not empty, rows are sequentially retrieved and assigned to the *$row* variable.

7. The row retrieved in *$row* variable is extracted to display data stored in the *image, title, author1,* and *price* columns.

8. A *quantity* input text field and two buttons, *Show Details* and *Add To Cart,* are attached to each book displayed.

9. The input text field *quantity* is assigned the default value of *1.* The user can change this value if desired.

10. The *Add To Cart* button invokes the *cart.php* PHP script file, which adds the selected book to the cart.

11. The *Show Details* button invokes the *showdetails()* function. The selected book's *isbn* is passed to *showdetails()* as a parameter. We will see in the next section how the *showdetails()* function is used for displaying the selected book's detailed information.

The response of the *getbooks.php* script is sent to the *selectedsubcat()* function, which appends the response to the *bookslist* div for displaying in the panel. The code for *booksdisplay div* is shown in Listing 6.11.

Listing 6.11. *booksdisplay* div

```
<div id="booksdisplay">
    <div class="toolbar">
        <a class="back" href="#">Back</a>
        <h1>Select Books</h1>
    </div>
    <div id="bookslist">
    </div>
</div>
```

The sequence of event is as follows:

1. The *booksdisplay* div contains a nested *toolbar* div that displays a *Back* button for jumping to the *Subcategories* panel.

2. The toolbar also contains a *Heading 1* element that displays the *Select Books* title on the panel. Below the toolbar is a *bookslist* div element.

3. The response generated by *getbooks.php,* containing all the books in the selected category and subcategory, is appended to *bookslist* via the *selectedsubcat()* function

6.5 Creating the *Book Details* Panel

Let's add the *Book Details* panel. The detailed information is fetched from the server-side *books* database table and displayed, as shown in Figure 6.5(a). The scrolled screen is shown in Figure 6.5(b).

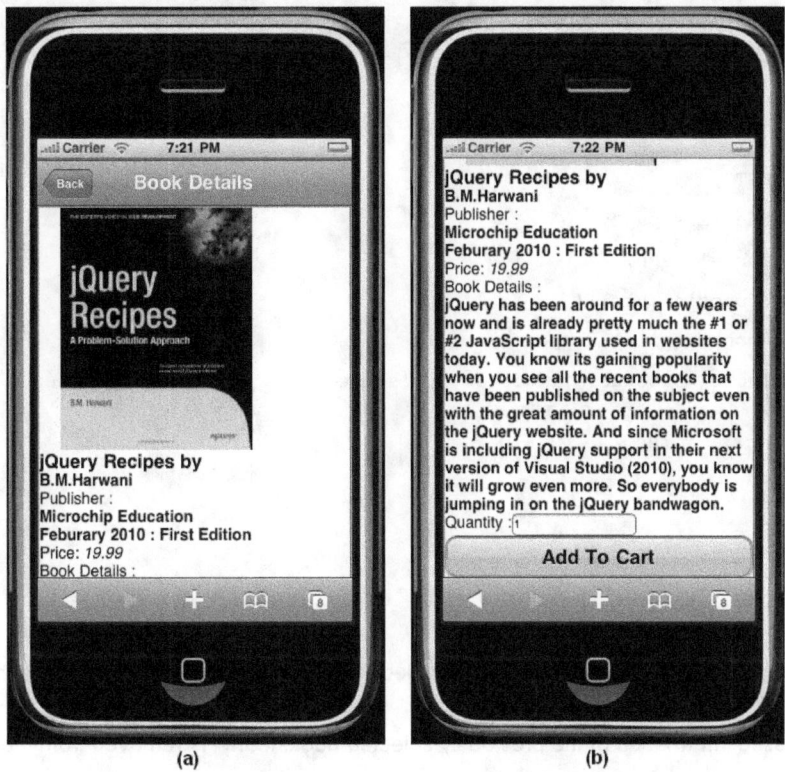

(a)   (b)

Figure 6.5. (a) Detailed information from the *Show Details* button

(b) More detailed information appears when the screen is scrolled

To display the detailed information of the selected book, we need the following:

- A *showdetails()* JavaScript function, which manages the entire process

- A *getdetails.php* PHP script that fetches from the *books* table the columns of the selected book containing the detailed information

- A *bookdetailsdisplay* div element, that displays the detailed information of the book

Let's first create the *showdetails()* JavaScript function, which performs the following tasks:

- Calls the *getdetails.php* server-side PHP script to fetch the selected book's columns from the server-side *books* database table containing detailed information, such as the publisher, coauthors, edition, publication date, and description

- Displays the detailed information fetched by the PHP script via the *bookdetailsdisplay* div.

- Navigates to the *bookdetailsdisplay* div, which outputs the data on the display

The *showdetails()* code is shown in Listing 6.12.

Listing 6.12. *showdetails()* function

```
function showdetails(isbn)
{
    $('#bookdetails').children().remove();
    $.ajax({
        type:"POST",
        url:"getdetails.php",
        data: 'isbn='+isbn,
        success:function(html){
            $('#bookdetails').append(html);
            jQT.goTo('#bookdetailsdisplay', 'slide');
        }
    });
    return false;
}
```

The sequence of events is as follows:

1. The *isbn* parameter of the book whose detailed information is desired is passed to the *showdetails()* function.

2. The detailed information of the previously selected book, if any, is removed from the *bookdetails* div.

3. The *ajax()* method is called with the POST request method. The request is sent to the server-side script file, *getdetails.php*.

4. The *isbn* number is passed as a parameter to *getdetails.php* in the form of *data*.

5. The script file fetches book details of the book with the specified ISBN from the *books* table.

6. When the request sent to *getdetails.php* succeeds, the *success* callback function is executed.

7. The script response is received as an *html* parameter.

8. The *html* parameter is appended to the *bookdetails* div element.

9. The *bookdetails* div element is nested inside a *bookdetailsdisplay* div element, hence we jump to the *bookdetailsdisplay* with a *slide* animation where the detailed book information (the response generated by the PHP script), is displayed.

10. To suppress the browser's default click-event behavior, the function returns *false*.

The code for server-side PHP script, *getdetails.php,* is shown in Listing 6.13.

110

Listing 6.13. *getdetails.php* script

```php
<?php
    $isbn =trim($_REQUEST['isbn']);
    $connect=mysql_connect("localhost","root", "mce") or die ("Please check your server
        connection");
    mysql_select_db("shopping");
    $query="Select isbn, title, author1, author2, author3, publisher,
        publish_date_edition, price, image, description from books where isbn ='$isbn'";
    $results =mysql_query($query) or die (mysql_query());
    while ($row=mysql_fetch_array($results))
    {
        extract ($row);
        echo '<fieldset style="background-color:white; color:black;">';
        echo '<form action="cart.php?isbn=' . $isbn . '&title=' . urlencode($title) .
            '&price=' . $price . '&action=add' . '" method="POST">';
        echo '<img src=' . $image .'>';
        echo '<h3>' . $title . ' by </h3>';
        echo '<h4>' . $author1 . '</h4>';
        if($author2 !='NULL')
            echo '<h4>' . $author2 . '</h4>';
        if($author3 !='NULL')
            echo '<h4>' . $author3 . '</h4>';
        echo '<label>Publisher :</label><h4>' . $publisher . '</h4>';
        echo '<h4>' . $publish_date_edition . '</h4>';
        echo '<label>Price: </label>';
        echo '<em>' . $price . '</em><br/>';
        echo '<label>Book Details :</label><h4>' . $description . '</h4>';
        echo '<label>Quantity :</label><input type="text" style="height:22px;"
            name="quantity" value="1" />';
        echo '<a class="submit whiteButton" href="#" onclick="this.form.submit();">
            Add To Cart</a>';
        echo '</form>';
        echo '</fieldset>';
    }
?>
```

The sequence of events is as follows:

1. The *isbn* of the requested book is passed to *getdetails.php*.

2. Using the *$_REQUEST* array, the *isbn* number is retrieved and stored in the variables *$isbn*.

3. A connection to the *shopping* database is established and a query executed to fetch detailed information from the *books* table for the specified *isbn*.

4. The *isbn, title, author1, author2, author3, publisher, publish_date_edition, price, image,* and *description* columns are fetched from the *books* table and stored in the *results* array.

5. The row containing information of the desired book is retrieved from the *results* array and assigned to *$row* variable.

6. The row retrieved in the *$row* variable is extracted, so the details can be displayed.

7. A *quantity* input field and an *Add To Cart* button is displayed.

8. The input field, *quantity,* has a default value of *1,* but the user can change it, if desired.

9. The *Add To Cart* button invokes the *cart.php* script file. Used to add the selected book to the cart .

10. The script response is sent to *showdetails()*, which appends the response to the *bookdetails* div.

The *bookdetails* div is nested inside the div, which outputs the information to the display. The code for *bookdetailsdisplay* is shown in Listing 6.14.

Listing 6.14. *bookdetailsdisplay* div

```
<div id="bookdetailsdisplay">

    <div class="toolbar">

        <a class="back"  href="#">Back</a>

        <h1>Book Details</h1>

    </div>

    <div id="bookdetails">

    </div>

</div>
```

The *bookdetailsdisplay* div contains a nested *toolbar* div with a *Back* button, used to return to the *Select Books* panel, as shown in Figure 6.4. The toolbar also contains a *Heading 1* element displaying the *Book Details* title. Below the toolbar is a *bookdetails* div element that displays, via the *showdetails()* function, the detailed information returned after a specific ISBN is passed to the *getdetails.php* script.

We are almost finished with the first phase of our *Book Store* application—just a few details remain.

6.6. Coding the Remaining *Home* Panel List Items

The most important list item in the *Home* panel is obviously *Books*, as it displays what the store has to offer and segues into the Shopping cart. But we have other list items on the Home panel as well—*Contact Us, New Arrivals, Discount Offers, Best Selling,* and *Gift Cards.* The code for these div elements is shown in Listing 6.15. The output was shown in Figure 1.19.

Listing 6.15. *contactus, newarrivals, discountoffers, bestselling,* and *giftcards* divs

```
<div id="contactus">

    <div class="toolbar">

        <h1>Contact Us</h1>
```

```html
        <a class="back" href="#">Back</a>
    </div>
    <div class="info">
        <p>XYZ Book Company</p>
        <p>11 Books Street, NY, NY 10012 </p>
        <p>USA</p>
        Email us: <a href="mailto:bmharwani@yahoo.com"
            target="_blank">bmharwani@yahoo.com</a>
    </div>
</div>


<div id="newarrivals">
    <div class="toolbar">
        <h1>New Arrivals</h1>
        <a class="back" href="#">Back</a>
    </div>
    <div class="info">
        <p>We have plenty of new arrivals</p>
        <p>New Books at exciting offers are available now</p>
    </div>
    <ul>
        <li>Linux for Lovers, by Bintu </li>
        <li>Master Unix Shell Programming, by Bintu </li>
        <li>Learn Knitting at Home, by Susan </li>
    </ul>
</div>


<div id="discountoffers">
    <div class="toolbar">
        <h1>Deep Discount</h1>
        <a class="back" href="#">Back</a>
    </div>
    <div class="info">
        <p>New Books at the deep discounts are available</p>
        <p>Discount is valid just for few days. So Hurry !!!</p>
    </div>
    <ul>
        <li>Introduction to MS-DOS 1.0, by Bintu </li>
```

```
            <li>Buggy Whip Construction, by Bintu </li>
        </ul>
</div>


<div id="bestselling">
    <div class="toolbar">
        <h1>Best Selling</h1>
        <a class="back" href="#">Back</a>
    </div>
    <div class="info">
        <p>Following is the list of the best selling books</p>
        <p>These books range from Computers to Story & Fiction</p>
    </div>
</div>


<div id="giftcards">
    <div class="toolbar">
        <h1>Gift Cards</h1>
        <a class="back" href="#">Back</a>
    </div>
    <div class="info">
        <p>Attractive Festive Gift Cards available at attractive prices</p>
        <p>Free shipping offers for few days</p>
    </div>
</div>
```

6.7 Summary

In this chapter, we developed all the panels required for the application. In the next chapter, we will proceed to the next step, maintaining the cart.

7
Assembling the Store, Part 2: Maintaining the Cart

In the previous chapter, we developed the first phase—Displaying Store Contents—and created the panels necessary to display the books for sale. It naturally follows that the user might want to make purchases. In this chapter, we'll create the code for cart maintenance, allowing users to add, update, and delete books from the shopping cart.

In this chapter, we will create two panels:

- The *Cart Updated* panel, which displays cart modifications and provides three navigation options:

  - The *Show Cart* link, which jumps to the *Items in Cart* panel.

  - The *Shopping* button which jumps to the *Categories* panels.

  - The *Check Out* button, which initiates order placement for cart items.

- The *Items in Cart* panel, which displays cart contents.

7.1 Creating the Cart Updated Panel

The Cart Updated panel, shown in Figure 7.1, is displayed when we perform any of the following operations:

- Adding a book to the cart
- Updating the quantity of any book in the cart
- Deleting a book from the cart

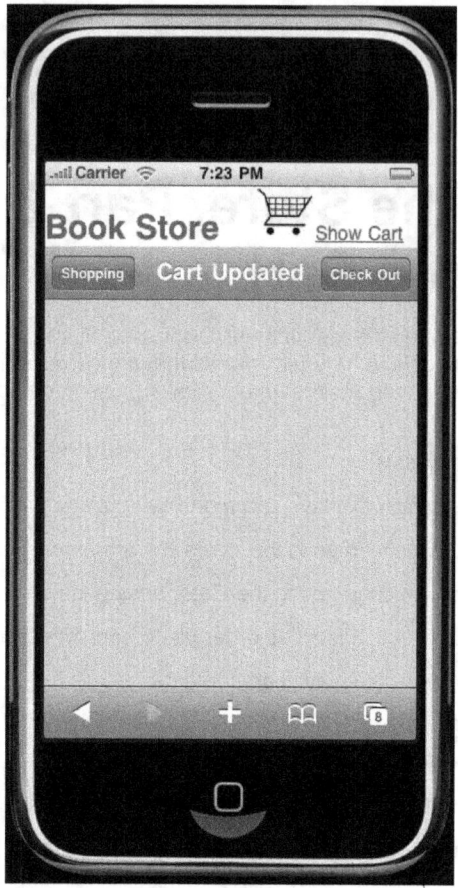

Figure 7.1. *Cart Updated* panel

The *Cart Updated* panel is displayed via the PHP script, *cart.php*, and its code is shown in Listing 7.1. But before we dissect the code, it's important to understand that this script is called from *getbooks.php*, *getdetails.php*, and the *dispcart()* function:

- Recall from Chapter 6, that *getbooks.php* displays the list of books in the selected category and subcategory via the *Select Books* panel. Two buttons, *Add To Cart* and *Show Details*, are attached to each book. When the *Add To Cart* button is selected, *cart.php* is called, and the selected book information—*isbn number, book title, quantity, price,* and *action*—is also passed to the script. The *action* parameter will be explained in more detail later. For now, just consider that the *action* value is set to *add* while adding a book to the cart.

- The *getdetails.php* file we saw in Chapter 6 displays the selected book's detailed information via the *Book Details* panel. The script is invoked when the *Show Details* button of any book is tapped. The detailed book information is followed by the *Add To Cart* button, which, when selected, invokes the *cart.php* script.

- The *dispcart()* function, explained in more detail later in this chapter, displays the cart contents. Every book being displayed has an input field, *quantity,* and two buttons, *Update* and *Delete*. When the *Update* button is selected, *cart.php* is called with the action parameter set to *update*. When the *Delete* button is selected, *cart.php* is called with the action parameter set to *delete*.

116

Listing 7.1. *cart.php* script

```php
<?php
    $isbn = trim($_REQUEST['isbn']);
    $title=$_REQUEST['title'];
    $qty=$_REQUEST['quantity'];
    $price=$_REQUEST['price'];
    $action=trim($_REQUEST['action']);
?>


<div id="cartupdated">
    <h1 style="display: table-cell; width: 190px;background-color:white; color: #blue;
        font: bold 28px Helvetica;"> Book Store </h1>
    <a style="display: table-cell; width: 150px; background-color:white;" href="#"
        onclick="dispcart(); jQT.goTo('#showcart', 'slide');">
        <img  src="cartfigure.tiff">Show Cart </a>
    <div class="toolbar">
         <h1>Cart Updated</h1>
        <a class="button leftButton" href="#" onclick="showcategories();"> Shopping</a>
        <a class="button" href="#" onclick="checkout();">Check Out</a>
    </div>
</div>


<script type="text/javascript">
    var sbn="<?php  echo $isbn; ?>";
    var tit="<?php echo $title; ?>";
    var qt="<?php echo $qty; ?>";
    var pr="<?php echo $price; ?>";
    var action="<?php echo $action; ?>";
    if(action=="add") addtocart(sbn,tit,pr,qt);
    if(action=="update")
    {
        if(qt>0) updateCart(sbn, qt);
        else deleteCart(sbn);
    }
    if(action=="delete")deleteCart(sbn);
</script>
```

This script not only displays the *Cart Updated* panel, but also acts as a branching zone that invokes the *addtocart(), updateCart(),* and *deleteCart()* functions, depending on the value of the *action* parameter. The sequence of events is as follows:

1. The script first retrieves the *isbn, title, quantity, price,* and *action* parameter values sent from the calling script or function, and stores them in the *$isbn, $title, $qty, $price* and *$action* variables respectively.

2. The value of the *action* parameter is set to *add, delete,* or *update,* depending on why the script was called.

3. The *Cart Updated* panel is created via the *cartupdated* div.

4. The panel has a header above the toolbar that displays *Book Store* in bold, blue 28-pixel Helvetica. Next to the header, is a *cart* icon and a *Show Cart* link. Selecting the *cart* icon or *Show Cart* link invokes the *dispcart()* function and jumps to the *showcart* div, which displays the *Items in Cart* panel.

5. A toolbar for the *Cart Updated* panel is created. It contains a header and two buttons: *Shopping* on the left side and *Check Out* on the right. Tapping the *Shopping* button invokes the *showcategories()* function, which jumps to the *Categories* panel. Tapping the *Check Out* button invokes the *checkout()* function.

6. The values in the *$isbn, $title, $qty, $price* and *$action* PHP variables are assigned to the JavaScript variables *isbn, tit, qt, pr* and *action.*

7. Depending on the *action* parameter value, the script invokes one of the three JavaScript functions: *addtocart(), updateCart(,)* and *deleteCart().* These functions do exactly what their names describe, th*at* is, adds a book to the cart, updates the cart, and deletes cart items.

7.2 Creating the Items in Cart Panel

The *Items in Cart* panel performs the following tasks:

- Jumps to the *Categories* panel via the *Shopping* button, so the user can continue shopping.

- Initiates the checkout process when the *Check Out* button is tapped.

- Updates the book quantity when the *Update* button is tapped.

- Deletes a book from the cart when the *Delete* button is tapped.

Every book displayed in the cart has a *quantity* input field and two buttons, *Delete* and *Update,* as shown in Figure 7.2(a). Let's add three copies of *Perfect Plants* to the cart. The result is shown in Figure 7.2(b).

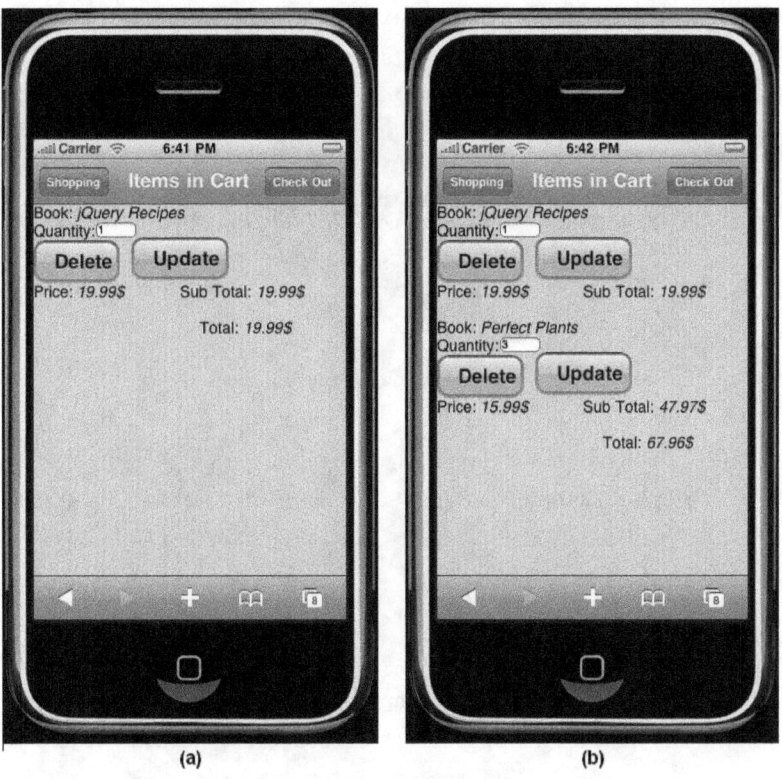

(a)   (b)

Figure 7.2. (a) Items inserted in the cart (b) Cart status

The code for displaying books in the *Items in Cart* panel is shown in Listing 7.2.

Listing 7.2. *dispcart()* function

```
function dispcart()
{
    var total=0;
    var subtot=0;
    var sid="<?php  echo $sessid; ?>";
    var datab;var shortName = 'tmpCart';
    var version = '1.0';
    var displayName = 'tmpCart';
    var maxSize = 200000;
    datab = openDatabase(shortName, version, displayName, maxSize);
    $('#showitems').children().remove();

    datab.transaction(
        function(transaction) {
            transaction.executeSql(
```

```
'SELECT cart_sess, cart_isbn,  cart_item_name, cart_qty, cart_price FROM
   shopcart where cart_sess=?;',[sid],
function (transaction, result) {
  if (result.rows.length <=0)
  {
     $('#showitems').append('<h1> The Cart is Empty </h1>');
  }
  else
  {
     for (var i=0; i < result.rows.length; i++) {
        subtot=0;
        var row = result.rows.item(i);
        $('#showitems').append('<div style="float:left;width:300px;">
           <label>Book: </label><em>' + row.cart_item_name + '</em>
           </div>');
        $('#showitems').append('<div style="float:left;width:200px;">
           <label style="float:left;" >Quantity: </label>
           <form action="cart.php?isbn='+row.cart_isbn+
             '&action=update'+'" method="POST" class="form">
              <input type="text" name="quantity"  size="6"
                 value="'+row.cart_qty+'"/>
              <a class="submit whiteButton" href="#"  style=" width:
                   55px;color:rgba(0,0,0,.9);float:right;">Update</a>
           </form>
           <form action="cart.php?isbn='+row.cart_isbn+
             '&action=delete'+'" method="POST" class="form">
              <a style="width: 45px;color:rgba(0,0,0,.9);
                   float:left;" href="#"  class="submit whiteButton">
                   Delete</a>
           </form>
           </div>');
        $('#showitems').append('<div style="float:left; width:150px;">
           Price: <em>' + row.cart_price + '$</em>');
        subtot=row.cart_qty*row.cart_price;
        $('#showitems').append('<div style="width: 150px;float:left;">
           Sub Total: <em>' + subtot.toFixed(2) + '$</em><br/><br/>');
        total=total+subtot;
     }
```

```
            $('#showitems').append('<div style="width: 150px;float:right;">
                Total: <em>' + total.toFixed(2) + '$</em>');
            }
        },
        displayerrormessage
    );
    }
);

return false;
}
```

The sequence of events is as follows:

1. Two variables, *total* and *subtot* are initialized to *0*.

2. The session id of the user stored in the *$sessid* PHP variable is assigned to the *sid* JavaScript variable. Recall that the user session id is created using the *session_start()* method and is uniquely assigned to each user. For more information, see the *index.php* source code in *Appendix C*.

3. The *subtot* variable totals the price of all the copies of a particular book by multiplying the *quantity* (copies) by the *price* of a single copy.

4. The *total* variable stores the sum of all the *subtot* variables—the total cost of books in the cart.

5. A connection to the client-side database, *tmpCart*, is established.

6. If the local database doesn't already exist by the specified name, a new one is created and the connection is stored in the *datab* object.

7. Any books shown earlier in the cart are removed from the *showitems* div.

8. An SQL SELECT statement is executed via the *transaction* object's *executeSql* method. This method finds the rows, if any, in the current user's *shopcart* table. The result of the SQL query is stored in a *result* array.

9. If the number of rows in the *result* array is <=0, there is no book selected in the cart, so we display a *The Cart is Empty* message in the panel by appending the message to the *showitems* div.

10. If the *result* array is not empty, we use a *for* loop to display each book or row in the *shopcart* table. To display books, we take one row at a time from the *result* array and store it in the *row* variable . The information in the *cart_item_name, cart_qty,* and *cart_price* columns within the *row* variable is displayed by appending it to the *showitems* div.

11. To allow the user to update a book quantity or delete the book from the cart, we display two buttons. To make them, we create two *submit whiteButton* hyperlinks, with the text *Update* and *Delete* respectively.

12. *Update* and *Delete* requests must be passed to the *cart.php* PHP script that was shown in Listing 7.1. To send the new quantity, we wrap the *quantity* input field inside a *form* element. Input elements within the *form* are automatically passed to the PHP script file named in the form's *action* attribute when the *Submit* button is tapped. For details, refer Chapter 3.

13. Two parameters are passed to *cart.php*: the book's *isbn* and the *action* parameter set to either *update* (if the *Update* button was pressed) or *delete* (if the *Delete* button is pressed).

14. The subtotal is computed by multiplying the *cart_qty* and *cart_price* fields. The result is stored in the *subtot* variable. The subtotal is displayed on the panel by appending it to the *showitems* div.

15. The subtotal of each book is added into the *total* variable.

16. The total of all the books in the cart is displayed by appending it to the *showitems* div.

Note: The subtotal and *total* are set to two decimal places by using the *toFixed()* method.

Information about cart contents is displayed via the *dispcart()* function by appending the data to the *showitems* div, which is nested inside the *showcart* div shown in listing 7.3.

Listing 7.3. *showcart* div

```
<div id="showcart">

   <div class="toolbar">

      <a class="button leftButton" href="#" onclick="showcategories();"> Shopping</a>

      <a class="button" href="#" onclick="checkout();">Check Out</a>

      <h1>Items in Cart</h1>

   </div>

   <div id="showitems" style="color:black;" >

   </div>

</div>
```

The *showcart* div contains a nested *toolbar* div with a *Shopping* button on the left and a *Check Out* button on the right. The *Shopping* button invokes the *showcategories()* function and the *Check Out* button initiates the checkout process, which we will cover in detail in Chapter 8. The toolbar also contains an *Items in Cart* Heading 1 element. Below the toolbar is a *showitems* div element used to display information about books in the cart.

7.2.1 Adding Items to the Cart

In our application, books are added to the cart when the *Add To Cart* button is tapped in the *Select Books* or *Book Details* panels. In response, *cart.php* is invoked with the action parameter set to *add. cart.php* will, in turn, invoke the *addtocart()* function, passing along information about the book being added to the cart, such as *isbn, title, price,* and *qty*.

The *addtocart()* function is shown in Listing 7.4.

Listing 7.4. *addtocart()* function

```
function addtocart(isbn, title, price,qty)

{

   if(qty >0)

   {

      var sid="<?php  echo $sessid; ?>";

      var datab;var shortName = 'tmpCart';

      var version = '1.0';

      var displayName = 'tmpCart';

      var maxSize = 200000;

      datab = openDatabase(shortName, version, displayName, maxSize);
```

```
datab.transaction(
    function(transaction) {
        transaction.executeSql(
            'CREATE TABLE  IF NOT EXISTS shopcart ' +
            ' (id INTEGER NOT NULL PRIMARY KEY AUTOINCREMENT, ' +
            ' cart_sess varchar(50), cart_isbn varchar(30),  cart_item_name
                varchar(100), cart_qty integer, cart_price float );'
        );
    }
);


datab.transaction(
    function(transaction) {
        transaction.executeSql(
            'SELECT cart_sess, cart_isbn,  cart_item_name, cart_qty, cart_price
                FROM shopcart where cart_sess=? and cart_isbn=?;',[sid, isbn],
            function (transaction, result) {
                if (result.rows.length >0)
                {
                    var row = result.rows.item(0);
                    qty=parseInt(qty)+parseInt(row.cart_qty);
                    datab.transaction(
                        function(transaction) {
                            transaction.executeSql(
                                'update shopcart set cart_qty=? where cart_sess=? and
                                    cart_isbn=?;',
                                [qty, sid, isbn],
                                function(){
                                },
                                displayerrormessage
                            );
                        }
                    );
                }
                else
                {
                    datab.transaction(
```

```
function(transaction) {
    transaction.executeSql(
        'INSERT INTO shopcart (cart_sess, cart_isbn,
            cart_item_name, cart_qty, cart_price) VALUES
                (?,?,?,?,?);',
        [sid, isbn, title, qty, price],
        function(){
        },
        displayerrormessage
    );
    }
    );
    }
    }
    );
    }
    );
    }
}
```

Adding a book to the cart means inserting a row into the client-side *tmpCart* database's *shopcart* table. The row to be inserted into *shopcart* includes the information such as the *isbn, title, price,* and *quantity* of the user-selected book. So, we must do the following:

- We first check that the quantity entered by the user is greater than 0, as a negative quantity cannot be added to cart.

- If the quantity of the book is greater than 0, a client-side database, called *tmpCart*, is created if it doesn't already exist. The connection is stored in a *datab* object.

- To create a *shopcart* table in the *tmpCart* database, we invoke *datab's transaction* method *and pass* an *anonymous function* to it.

- We pass the *transaction* to the *anonymous function* and then execute an SQL CREATE statement via the *executeSql* method (for details, refer to Chapter 4). The *shopcart* table has six columns:

  - *id*—A primary key set to AUTOINCREMENT so that it stores a unique value for every row inserted into the table. The value is automatically incremented by 1 for every new row added.

  - *cart_sess*—Stores the user's session id

  - *cart_isbn*—Stores the selected book's ISBN number

  - *cart_item_name*—Store's the book's title

  - *cart_qty*—Stores the book's quantity

  - *cart_price*—Stores the book's price

- If the user selects the same book item again, this new quantity is added to the quantity already in the cart. For example, if cart contains one copy of a book and the user selects the same book again from the *Select Books* panel with the quantity field set to 2, then the total number of that book in the cart will be updated to 3. To implement this feature, we need to first check, via an SQL SELECT statement, to see if a book with the same *isbn* already exists in the cart before adding a new row to the *shopcart* table.

- The result of the query is stored in the *result* array . If the row value in the *result* array is greater than 0, a book with the same *isbn* already exists in the *shopcart* table, so we retrieve the row from the *result* array and store it in the *row* variable.

- The value in the *cart_qty* column, which is the number of copies in the cart, is added to any additional quantity entered by the user.

- If an error occurs while the SQL statement is executed, the *displayerrormessage* function is invoked to display an error message.

- If the *result* array has no rows, it means no book with the same ISBN is in the *shopcart*. In that case, we execute an SQL INSERT statement to insert a new row in the *shopcart* table. The row contains information passed via the parameters *isbn, title, price,* and *qty.*

Writing the error-handling function is trivial—the code is shown in Listing 7.5.

Listing 7.5. displayerrormessage() Function

```
function displayerrormessage(transaction, error) {
    alert('Error:  '+error.message+' has occurred with Code: '+error.code);
    return true;
}
```

The function displays the *error message* and *error code* of the goof-up. The function returns *true* to halt execution and rolls back the entire transaction.

Note: If the error handling function returns *false*, the transaction will continue.

7.2.2 Updating Items in the Cart

The cart contents are displayed in the *Items in Cart* panel via the *dispcart()* function. Besides displaying book information, the panel also contains buttons to update the quantity or delete the book from the cart.

Let's update the number of copies of *jQuery Recipes,* shown in Figure 7.2(b), from one to two. Once we enter the new value, we are sent to the *Cart Updated* panel shown in Figure 7.1. We can see the modified cart by tapping either the *Showcart* link or *cart* icon, as shown in Figure 7.3.

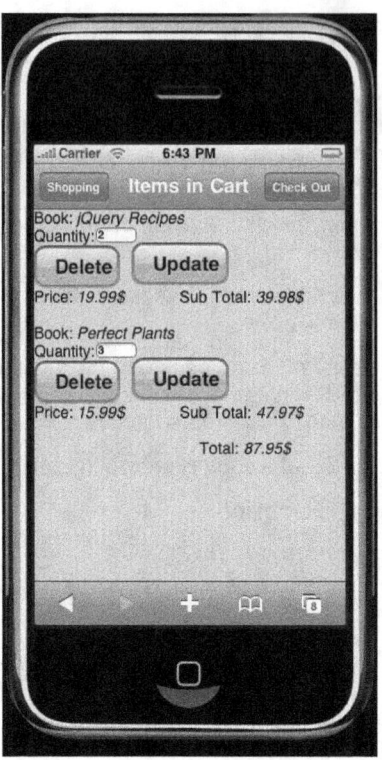

Figure 7.3. Cart status after updating a book quantity

The code of the *updateCart()* function is shown in Listing 7.6.

Listing 7.6. *updateCart()* function

```
function updateCart(isb, qty)
{
    var sid="<?php  echo $sessid; ?>";
    var datab;var shortName = 'tmpCart';
    var version = '1.0';
    var displayName = 'tmpCart';
    var maxSize = 200000;
    datab = openDatabase(shortName, version, displayName, maxSize);

    datab.transaction(
        function(transaction) {
            transaction.executeSql(
                'update shopcart set cart_qty=? where cart_sess=? and cart_isbn=?;',
                [qty, sid, isb],
                function(){
                },
```

```
            displayerrormessage
        );
    }
  );

  return false;
}
```

The sequence of events is as follows:

1. The *updateCart()* function is invoked when user changes the *quantity value* and selects the *Update* button.

2. Two parameters are passed to the function: *isbn* and *qty*. The former is the book's ISBN number and the latter is the new *quantity*.

3. We need to update the *shopcart* table of the *tmpCart* client-side database. A connection to *tmpCart* is established and the connection is stored in a *datab* object.

4. An SQL UPDATE statement is executed to update the *cart_qty* column value to the new user-entered quantity.

7.2.3 Deleting Items from the Cart

The *dispcart()* function allows us to update a book quantity, as well as delete a book from the cart. For example, Figure 7.3 showed two books in the cart: *jQuery Recipes* and *Perfect Plants*. If we remove *jQuery Recipes* by tapping its associated *Delete* button, we'll see the *Cart Updated* panel (Figure 7.1). From here, we can tap the *Show Cart* link or the *cart* icon and see the updated cart contents, as shown in Figure 7.4(a). If we delete the remaining book, we'll see the empty cart message shown in figure 7.4(b).

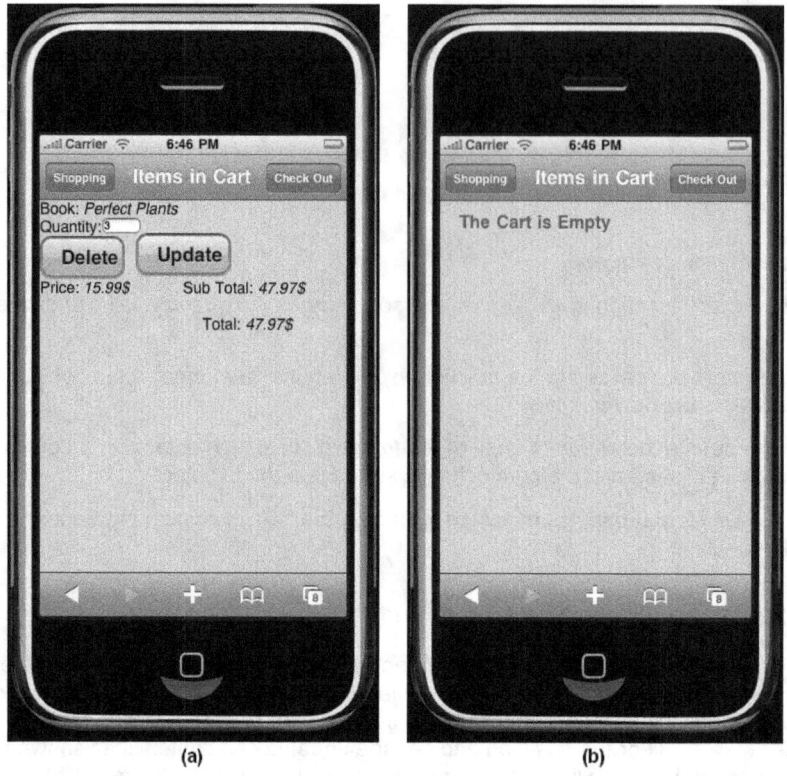

(a)   (b)

Figure 7.4. (a) The cart contents after a book deletion (b) Empty cart message

The books from the cart are deleted by calling the *deleteCart()* function from the PHP script, *cart.php (*see Listing 7.1), and passing the *isbn* parameter. The code of the d*eleteCart()* function is shown in Listing 7.7.

Listing 7.7. *deleteCart()* function

```
function deleteCart(isb)
{
    var sid="<?php  echo $sessid; ?>";
    var datab;var shortName = 'tmpCart';
    var version = '1.0';
    var displayName = 'tmpCart';
    var maxSize = 200000;
    datab = openDatabase(shortName, version, displayName, maxSize);

    datab.transaction(
        function(transaction) {
            transaction.executeSql(
                'delete from shopcart where cart_sess=? and cart_isbn=?;',
                [sid, isb],
```

```
        function(){
        },
        displayerrormessage
    );
    }
);
    return false;
}
```

The sequence of events is as follows:

1. The *deleteCart()* function is invoked when the user selects any book's *Delete* button in the *Items in Cart* panel.

2. The *Items in Cart* panel is displayed through the *dispcart()* function (see Listing 7.2). The *cart.php* script is executed with the *action* parameter set to *delete*. This, in turn, invokes the *deleteCart()* function, which performs the deletion.

3. The *isbn* of the book to be deleted is supplied to *deleteCart()* and stored in the *isbn* parameter.

4. To delete a book from the cart means deleting it from *tmpCart*'s *shopcart* table. A connection to *tmpCart* is established as usual and the connection is stored in a *datab* object.

5. An SQL DELETE statement is executed to delete the book from the *shopcart* table whose *cart_isbn* column matches the *isbn* parameter and whose *cart_sess* matches the user's session id.

7.3 Summary

In this chapter, we created the code required to maintain the cart. Now lets go ahead and develop the last phase of our application: placing an order for the cart contents.

8

Assembling the Store, Part 3: Placing Orders

In the previous chapter, we developed code for maintaining the bookstore cart. In this chapter, we'll discuss the third and final phase—creating accounts, logging in, and supplying shipping information. We will be creating following panels:

- *Checking Out* panel—Checks to see if the user is already logged in and provides choices: Supply Shipping Info, Sign In, or Create a New Account depending on the app's status.

- *Create Account* panel—Allows the user to create a new account.

- *Sign In* panel—Allows user authentication.

- *Placing Order* panel—Allows the user to supply shipping information.

- *Thank You* panel—Displays a thanks message and the order number for future reference.

Let's begin with creation of the *Checking Out* panel

8.1 Creating the *Checking Out* Panel

The *Checking Out* panel, shown in Figure 8.1, is displayed when we tap the *Check Out* button in the *Cart Updated* panel (see Figure 7.1) or the *Items in Cart* panel (see Figure 7.2). The panel checks to see if the user is logged in. If not, the user is provided with two buttons: *Sign In* and *Create Account*.

Figure 8.1. *Checking Out* panel, allowing users to either sign in or create a new account

Selecting S*ign In* jumps to the *Sign In* panel shown in Figure 8.4(a). From here, the user can sign into the application by entering a user ID and password. Selecting *Create Account* jumps to the *Create Account* panel (see Figure 8.2) allowing the user to create a new account.

To create the *Checking Out* panel, we need to build the following:

> *checkout()* function—Checks to see if there is anything in the cart and, depending on the result, either displays *Cart is empty* via the *checkout* div or branches to the *checkout.php* script to see if the user is logged in.

> *checkout* div—Displays the *checkout()* function messages and the *checkout.php* responses.

> *checkout.php*—Checks to see if the user is logged in. If so, the script prompts for shipping information. If not, the script prompts the user to either sign in or create a new account.

The code for the *checkout()* function is shown in Listing 8.1.

Listing 8.1. *checkout()* function

```
function checkout()
{
    $('#checkinfo').children().remove();
    var sid="<?php  echo $sessid; ?>";
    var datab;var shortName = 'tmpCart';
```

```
var version = '1.0';
var displayName = 'tmpCart';
var maxSize = 200000;
datab = openDatabase(shortName, version, displayName, maxSize);
datab.transaction(
    function(transaction) {
        transaction.executeSql(
            'SELECT cart_sess, cart_isbn,  cart_item_name, cart_qty, cart_price FROM
                shopcart where cart_sess=?;',[sid],
            function (transaction, result) {
                if(result.rows.length >=1)
                {
                    $.ajax({
                        type:"POST",
                        url:"checkout.php",
                        success:function(html){
                            $('#checkinfo').append(html);
                        }
                    });
                }
                else
                {
                    $('#checkinfo').append('<h1> Cart is empty</h1>');
                }
            },
            displayerrormessage
        );
    }
);
jQT.goTo('#checkout', 'slide');
return false;
}
```

The sequence of events is as follows:

1. Earlier messages displayed in the *checkinfo* div from previous checkout attempts are deleted.
2. The user's session id is obtained and stored in the *sid* variable.
3. The *tmpCart* database is opened and the connection stored in a *datab* object.

132

4. The *shopcart* table of the *tmpCart* database is searched with an SQL SELECT query to see if there is an item in the cart.

5. The result of the SQL SELECT query is stored in a *result* array.

6. The *result* array is checked to see if it has rows. If there are no rows, it means the cart is empty and it's obviously pointless to continue check out. A *Cart is empty* message is displayed on the panel.

7. If there are rows in the *result* array, the *checkout.php* script is invoked, which checks to see if the user is logged in. If so, the user is asked to provide shipping information. If not, the two buttons shown in Figure 8.1 are displayed, allowing the user to either sign in or create a new account.

The messages from *checkout()* and the response from *checkout.php* are displayed via the *checkout* div, so let's create it. The code appears in Listing 8.2.

Listing 8.2. *checkout* div

```
<div id="checkout">
    <div class="toolbar">
        <a  class="cancel" href="#">Cancel</a>
        <h1>Checking Out</h1>
    </div>
    <div id="checkinfo" >
    </div>
</div>
```

The *checkout* div has a nested *toolbar* div with a *Cancel* button to quit the check out process and return to the previous panel. It also has a *heading 1* element that displays the *Checking Out* panel title. Below the *toolbar* div is the *checkinfo* div, which displays the *checkout()* function message and the *checkout.php* response.

The code for *checkout.php* appears in Listing 8.3.

Listing 8.3. *checkout.php* script

```
<?php
    $uid=$_SESSION['userid'];
    $pwd=$_SESSION['password'];
    echo '<ul class="rounded">';
    if ((isset($_SESSION['userid']) && $_SESSION['userid'] != "") ||
        (isset($_SESSION['password']) && $_SESSION['password'] != ""))
    {
        echo'<li>If you are over with Shopping, please provide shipping information</li>';
        echo '<li><a href="#" class="submit whiteButton" onclick="shippinginfo(\'' . $uid .
            '\');">Supply Shipping Info</a></li>';
    }
```

```
    else
    {
        echo'<li>You are not Signed in yet</li>';
        echo '<li><a href="#" class="submit whiteButton" onclick="loginform();">Sign
            In</a></li>';
        echo '<li><a href="#" class="submit whiteButton" onclick="registrationform();">
            Create Account</a></li>';
    }
    echo '</ul>';
?>
```

The sequence of events is as follows:

1. We first check to see that the user ID and password are set in the *$_SESSION* array, which indicates that the user is logged in.

2. If user is logged in, he or she is asked to select the *Supply Shipping Info* button.

3. If the user is not logged in, a *You are not signed in yet* message is displayed, along with the *Sign In* and *Create Account* buttons.

4. If the *Sign In* button is selected, a *loginform()* function is invoked. This function displays a panel to enter a user ID and password. The *Create Account* button invokes the *registrationform()* function, allowing the user to create a new account.

8.2 Creating the *Create Account* Panel

The *Create Account* panel is displayed when a user selects the *Create Account* button from the *Checking Out* panel. The panel has input fields for creating a new account, as shown in Figure 8.2(a). Because the fields can't fit on a single screen, the user will have to scroll down to see the rest of the form, as shown in Figure 8.2(b). After the form is complete, tapping the *Submit* button creates the account. Validation checks are applied to the fields, Figure 8.2(c) shows what happens if the User ID field is left blank.

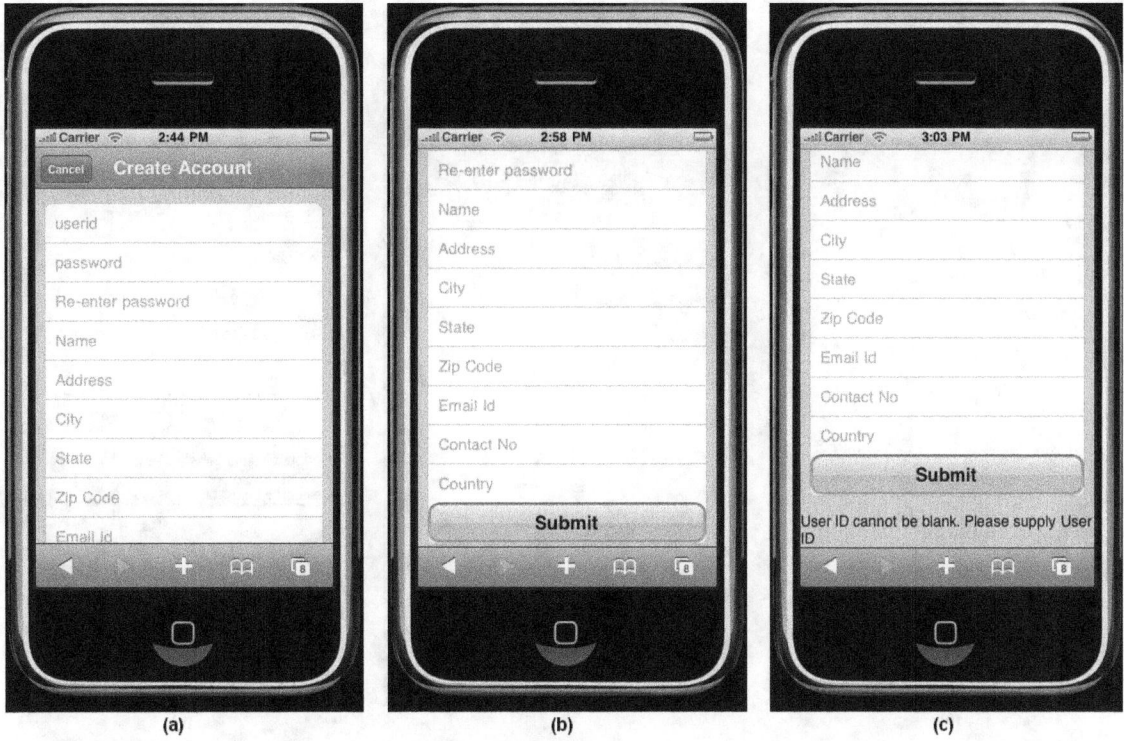

Figure 8.2. (a) Top half of the *Create Account* screen (b) Bottom half of the *Create Account* screen (c) Error message displayed if *User Id* is blank

Similarly, if the *password* and *Re-enter password* fields don't match, another error message is displayed, as shown in Figure 8.3(a). If all the data is entered correctly, as shown in Figure 8.3(b), a new account is created and a *Congratulations* message is displayed. The user is then prompted to *Sign In*, as shown in Figure 8.3(c).

(a)   (b)   (c)

Figure 8.3. (a) Error message displayed when the passwords don't match (b) Correctly entered user information (c) Congratulations message displayed after successful new account creation

To build the *Create Account* panel, we need the following:

- **registrationform() function**—Navigates to *createacct* div, which displays input fields required for account creation.

- **createaccount.php**—Creates a *createacct* div, which displays the input fields required for creating an account, as shown in Figure 8.2(a). The *createacct* div has a nested div, *userdetails,* that displays invalid data error messages from *createaccount()*. This div also displays responses generated from the *savecustomer.php* PHP script.

- **createaccount() function**—Performs validation checks on data entered in the *Create Account* panel. It also passes the data to the *savecustomer.php* PHP script, which save the information in the server-side *customers* database table.

- **savecustomer.php script**—Receives the data sent by *createaccount()* and inserts a new row in the *customers* table of the *shopping* database, which creates a new account. The script also displays a *Welcome message* and prompts the user to *Sign In.*

Let's start creating *Create Account* panel by coding the *registrationform()* function invoked when the *Create Account* button is selected from the *Checking Out* panel. The code of *registrationform()* is shown in Listing 8.4.

Listing 8.4. *registrationform()* function

```
function registrationform()
{
    jQT.goTo('#createacct', 'slide');
}
```

This function jumps to the *createacct* div, which displays the new account input fields.. The *createacct* div code, shown in Listing 8.5, could have been written into *index.php*, but because the code is somewhat lengthly, we'll create a separate file instead.

Listing 8.5. *createaccount.php* script

```
<div id="createacct">
    <div class="toolbar">
        <h1>Create Account</h1>
        <a class="button cancel" href="#">Cancel</a>
    </div>
    <ul class="rounded">
        <li><input type="text" id="user" placeholder="userid" /></li>
        <li><input type="password" id="passwd" placeholder="password" /></li>
        <li><input type="password" id="confirmpass" placeholder="Re-enter password" />
            </li>
        <li><input type="text"  id="name" placeholder="Name" /></li>
        <li><input type="text"  id="address" placeholder="Address" /></li>
        <li><input type="text"  id="city" placeholder="City" /></li>
        <li><input type="text"  id="state" placeholder="State" /></li>
        <li><input type="text"  id="zipcode" placeholder="Zip Code" /></li>
        <li><input type="text"  id="emailid" placeholder="Email Id" /></li>
        <li><input type="text" id="contactno" placeholder="Contact No" /></li>
        <li><input type="text" id="country" placeholder="Country" /></li>
        <a href="#" class="submit whiteButton" onclick="createaccount();">Submit</a>
    </ul>
    <div id="userdetails" style="color:black;">
    </div>
</div>
```

The sequence of events is as follows:

1. The *createacct* div creates the panel shown in Figure 8.2(a). It contains a nested *toolbar* div that displays a *Cancel* button and the *Create Account* panel title. The *Cancel* button cancels the create account operation return to the previous panel.

2. Below the *toolbar* div is an unordered list consisting of several list items. Wrapped inside the list items are the input fields that allow the user to enter information. The fields are *userid, password, Re-enter password, name, address, city, state, zip code, email id, contact number,* and *country*. For the purpose of retrieving data entered in the input fields, each of these is assigned a unique id: *user, passwd, confirmpass, name, address, city, state, zipcode, emailid, contactno,* and *country* respectively.

3. Below the input text fields is a *Submit* button that invokes the *createaccount()* function. In the *createaccount()* function, the data entered in the input fields is validated and passed to *savecustomer.php* for inserting a new row in the *customers* table of the *shopping* database.

4. The unordered list is followed by a *userdetails* div, which displays the response generated by *savecustomer.php*.

Let's now write the code for the the *createaccount()* function, as shown in Listing 8.6.

Listing 8.6. *createaccount()* function

```
function createaccount()
{
    var usr=$('#user').val();
    var pswd=$('#passwd').val();
    var cfpswd=$('#confirmpass').val();
    var name=$('#name').val();
    var add=$('#address').val();
    var city=$('#city').val();
    var state=$('#state').val();
    var zip=$('#zipcode').val();
    var email=$('#emailid').val();
    var contact=$('#contactno').val();
    var country=$('#country').val();

    $('#userdetails').children().remove();
    if(usr.length <=0)
    {
        $('#userdetails').append('<p> User ID cannot be blank. Please supply User ID
            </p>');
        return;
    }
    if(pswd.length <=0)
    {
        $('#userdetails').append('<p> Password cannot be blank. Please supply password
            </p>');
```

```
        return;
    }
    if(pswd !=cfpswd)
    {
        $('#userdetails').append('<p> Password and Re-enter password don\'t match. Please
            enter again </p>');
        return;
    }
    if(name.length <=0)
    {
        $('#userdetails').append('<p> Name cannot be blank. Please enter your name </p>');
        return;
    }
    if(contactno.length <=0)
    {
        $('#userdetails').append('<p> Contact Number cannot be blank. Please supply
            contact number </p>');
        return;
    }
    if(emailid.length <=0)
    {
        $('#userdetails').append('<p> Email Address cannot be blank. Please supply email
            address </p>');
        return;
    }
    var data='userid='+usr+'&password='+pswd+'&name='+name+'&add='+add+'&city='+city+
    '&state='+state+'&zip='+zip+'&email='+email+'&contact='+contact+'&country='+country;
    $('#userdetails').children().remove();
    $.ajax({
        type:"POST",
        url:"savecustomer.php",
        data: data,
        success:function(html){
            $('#userdetails').append(html);
        }
    });
    return false;
}
```

The sequence of events is as follows:

1. Error messages will be displayed by appending them to the *userdetails* div. We can see that the data entered into the *user, passwd, confirmpass, name, address, city, state, zipcode, emailid, contactno,* and *country* input fields is retrieved and stored temporarily in the *usr, pswd, cfpswd, name, add, city, state, zip, email, contact,* and *country* variables respectively.

2. Any previous error messages displayed via *userdetails* div is removed.

3. The length of the stored variables is checked to see if any are less than or equal to 0, indicating that the field is blank. The appropriate error message is displayed via *userdetails* div.

4. The data entered in the *pswd* and *cfpswd* variables are compared to see if they match. If their contents don't match, an error message is displayed.

5. The validated data is passed to *savecustomer.php* for inserting into the server-side database table *customers*.

6. The response from *savecustomer.php* is displayed by appending it to the *userdetails* div.

The new account is actually created when a new row containing information of the new user is inserted into the *customers* table. This process is performed by the *savecustomer.php* PHP script, so let's create it. The code is shown in Listing 8.7.

Listing 8.7. *savecustomer.php* script

```php
<?php
    $uid =trim($_REQUEST['userid']);
    $pswd =trim($_REQUEST['password']);
    $name =trim($_REQUEST['name']);
    $add =trim($_REQUEST['add']);
    $city =trim($_REQUEST['city']);
    $state =trim($_REQUEST['state']);
    $zip =trim($_REQUEST['zip']);
    $email =trim($_REQUEST['email']);
    $contact =trim($_REQUEST['contact']);
    $country =trim($_REQUEST['country']);
    $connect=mysql_connect("localhost","root", "mce") or die ("Please check your server
        connection");
    mysql_select_db("shopping");
    $query="INSERT INTO customers (userid, password, name, address, city, state, zipcode,
        emailid, contact_no, country) VALUES('$uid', '$pswd', '$name', '$add', '$city',
        '$state', '$zip', '$email','$contact', '$country')";
    $results =mysql_query($query);
    echo '<p>Congratulations '  . $uid . '!. Your account is created </p>';
    echo '<p>Select the below button to Sign In </p>';
    echo '<a href="#" class="submit whiteButton" onclick="loginform();">Sign In</a>';
?>
```

The sequence of events is as follows:

1. Data entered by the user in the *Create Account* panel is sent to the script via the *createaccount()* function.

2. The new data is fetched via the *$_REQUEST* array and stored in the *$uid, $pswd, $name, $add, $city, $state, $zip, $email, $contact,* and *$country* variables, respectively.

3. The connection to the *shopping* database is established.

4. An SQL INSERT statement is created and executed to store the new customer information in the *customers* database table.

5. After successful execution of the SQL statement, a *Congratulations message* is displayed, along with a *Sign In* button asking user to sign into the web application.

6. The response of the above script, that is, the *Congratulations message* and *Sign In* button, are returned to the *createaccount()* function and assigned to the *html* parameter of the *success* callback function.

7. The contents of the *html* parameter are displayed via *userdetails* div.

Lets now create a *Sign In* panel that authenticates the user.

8.3 Creating the *Sign In* Panel

The *Sign In* panel contains two input fields, *Userid* and *Password,* as shown in Figure 8.4(a). Validation checks are applied to both the input fields. If the user leaves the *Userid* field blank and taps the *Sign In* button, a *User id cannot be blank. Please supply userid* message is displayed, as shown in Figure 8.4(b). Similarly, if Password is blank, a *Password cannot be blank. Please supply password* error message is displayed, as shown in Figure 8.4(c).

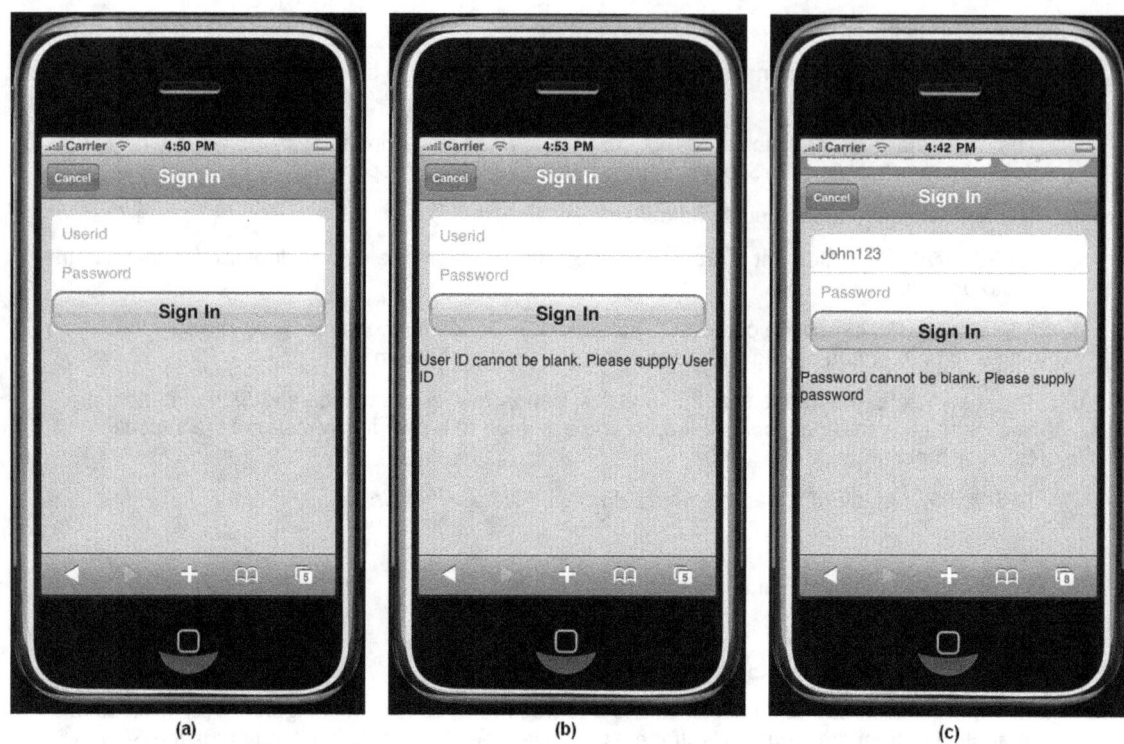

(a)   (b)   (c)

Figure 8.4. (a) *Sign In* panel (b) Error message displayed if *userid* field is blank

(c) Error message displayed if *password* field is blank

If the fields are filled in but incorrect, the error message *Sorry the User ID or Password is incorrect* is displayed, as shown in Figure 8.5(a). Two buttons, *Try Again* and *Create Account* are shown so the user can either try again or create a new account, if necessary. The *Try Again* button removes previously displayed error messages and allows the user to re-enter a user ID and password. The *Create Account* button will open the *Create Account* panel (see Figure 8.2(a), allowing user to create a new account.

A *welcome* message is displayed when the correct user ID and password have been entered, followed by a *Supply Shipping Info* button, as shown in Figure 8.5 (b).

142

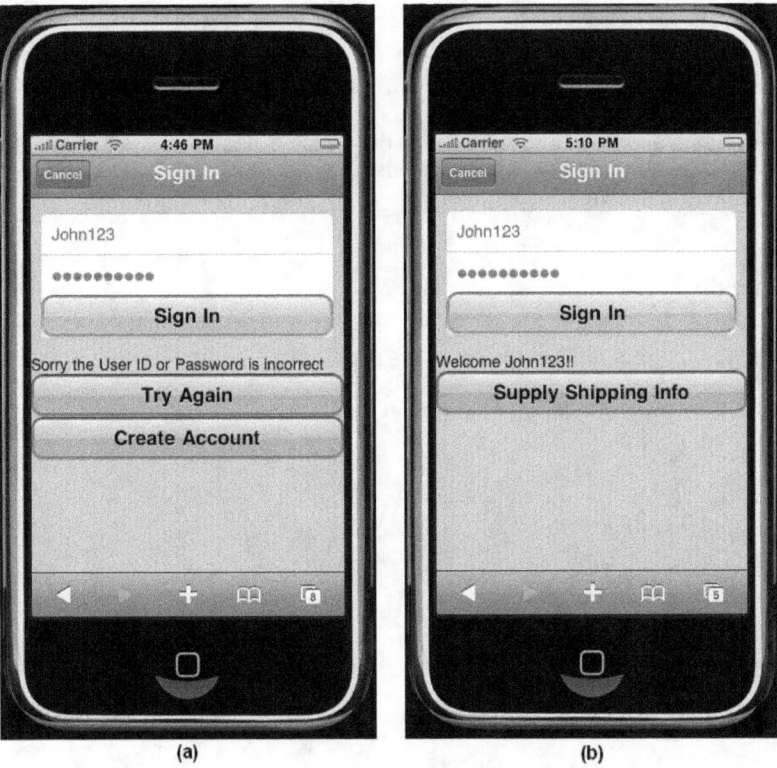

(a)   (b)

Figure 8.5. (a) Error message and buttons displayed if the user ID or password are incorrect. (b) Welcome message displayed if the user ID and password are correct

To create the *Sign In* panel, we have to build the following:

- **loginform() function**—Jumps to the *signin* div, which displays the input fields required for authenticating the user.

- **signin div**—Displays the *userid* and *password* input fields, followed by a *Sign In* button.

- **checklogin() function**—Validates *userid* and *password* and passes the validated data to *validatelogin.php*.

- **validatelogin.php script**—Compares the new data with that stored in the server-side database tables and displays the appropriate response.

- **trylogin() function**—Removes any previous messages and allows user to re-enter the user ID and password.

Let's begin creating *Sign In* panel by coding the *loginform()* function, shown in Listing 8.8.

Listing 8.8. *loginform()* function

```
function loginform()
{
    $('#response').children().remove();
```

```
        jQT.goTo('#signin', 'slide');

}
```

First, previous messages are removed from *response* div, which is nested inside the *signin* div. The code then jumps the *signin* div, which displays the input fields for the user ID and password, as shown in Figure 8.4(a).The *signin* div code is shown in Listing 8.9.

Listing 8.9. *signin* div

```
<div id="signin">

    <div class="toolbar">

        <a class="button cancel" href="#">Cancel</a>

        <h1>Sign In</h1>

    </div>

    <ul class="rounded">

        <li><input type="text"  id="userid" placeholder="Userid" /></li>

        <li><input type="password"  id="password" placeholder="Password" /></li>

        <a href="#" class="submit whiteButton" onclick="checklogin();">Sign In</a>

    </ul>

    <div id="response">

    </div>

</div>
```

The sequence of events is as follows:

1. The *signin* div contains a nested toolbar div with a *Cancel* button that quits the current operation and returns to the previous panel. The toolbar also contains the *heading 1* element that displays the *Sign In* title on the panel.

2. Following the *toolbar* div is an unordered list containing two list items. Wrapped inside the list items are the *userid* and *password* input fields. Unique IDs are assigned to the input fields so that data entered into them can be fetched.

3. Below the input fields is a *Sign In* button which, when selected, invokes the *checklogin()* function. This function validates the data and forwards it to *validatelogin.php* for authentication.

4. Following the unordered list is a *response* div where the response of the validation performed by *checklogin()* and *validatelogin.php* is displayed.

The code for the *checklogin()* function is shown in Listing 8.10.

Listing 8.10. *checklogin()* function

```
function checklogin()
{
    var usr=$('#userid').val();
```

```
var pswd=$('#password').val();
$('#response').children().remove();
if(usr.length <=0)
{
    $('#response').append('<p> User ID cannot be blank. Please supply User ID </p>');
    return;
}
if(pswd.length <=0)
{
    $('#response').append('<p> Password cannot be blank. Please supply password
        </p>');
    return;
}
$('#response').children().remove();
$.ajax({
    type:"POST",
    url:"validatelogin.php",
    data: 'userid='+usr+'&password='+pswd,
    success:function(html){
        $('#response').append(html);
    }
});
return false;
}
```

The sequence of events is as follows:

1. The user ID and password entered into the *Sign In* panel's input fields are retrieved and stored in the JavaScript variables *usr* and *pswd* respectively.

2. The *userid* and *password* input fields are identified through the IDs assigned to them.

3. The lengths of the *usr* and *pswd* variables are checked to see if either of them is less than or equal to 0. If either field is left blank, the respective variable's length will be 0, and a corresponding error message is displayed.

4. The error message is displayed by appending it to the *response* div.

5. Before displaying the error message, any earlier messages, if any, are first removed to avoid confusion.

6. On passing the validation checks, *userid* and *password* are passed to *validatelogin.php,* which compares the new data to that stored in customers table.

The code of *validatelogin.php* is shown in Listing 8.11.

Listing 8.11. *validatelogin.php* script

```php
<?php
    $uid =trim($_REQUEST['userid']);
    $pswd =trim($_REQUEST['password']);
    $connect=mysql_connect("localhost","root", "mce") or die ("Please check your server
        connection");
    mysql_select_db("shopping");
    $query="Select userid, password from customers where userid ='$uid' and password =
        '$pswd'";
    $results =mysql_query($query) or die (mysql_query());
    if(mysql_num_rows($results)>0)
    {
        echo 'Welcome ' . $uid . '!!';
        echo '<a href="#" class="submit whiteButton" onclick="shippinginfo(\'' . $uid .
            '\');">Supply Shipping Info</a>';
    }
    else
    {
        echo '<p>Sorry the User ID or Password is incorrect</p>';
        echo '<a href="#" class="submit whiteButton" onclick="trylogin();">Try Again</a>';
        echo '<a href="#" class="submit whiteButton" onclick="registrationform();">Create
            Account</a>';
    }
?>
```

The sequence of events is as follows:

1. The *userid* and *password* are passed to *validatelogin.php* via the *checklogin()* function.

2. The script fetches them via a *$_REQUEST* array and stores them in the *$uid* and *$pswd* variables respectively.

3. The connection to the *shopping* database is established.

4. An SQL SELECT statement is executed to see if a row in the *customers* table matches the *userid* and *password* specified in the *$uid* and *$pswd* variables.

5. If there is a row that matches, the supplied *userid* and *password* are correct and a *Welcome* message is displayed, along with a *Supply Shipping Info* button.

6. When the *Supply Shipping Info* button is tapped, the *shippinginfo()* function is invoked and the *userid* stored in the *$uid* variable is passed to it.

7. If no row matches, a *Sorry the User ID or Password is incorrect* message is displayed on the panel. Below the *Sorry* message, two buttons are displayed: *Try Again* and *Create Account*. The former

invokes the *trylogin()* function and the latter invokes the *registrationform()* function. The two functions allows the user to retry the authentication procedure or create a new account.

The code of the *trylogin()* function appears as shown in Listing 8.12

Listing 8.12. *trylogin()* function

```
function trylogin()
{
    $('#response').children().remove();
}
```

This simple function removes previous messages from the *response* div. The program remains on the *Sign In* panel, allowing user to re-enter the user ID and password.

Once the user is authenticated, the only remaining task left is supplying shipping information.

8.4 Creating the *Placing Order* Panel

The *Placing Order* panel displays user information that already exists in the *customers* table, as shown in Figure 8.6(a). This data is read-only and cannot be edited. Below this data are shipping data fields that must be completed, as shown in Figure 8.6(b).

Figure 8.6. (a) Supplying shipping information (b) Fields with completed shipping information

147

To create the *Placing Order* panel, we have to build the following :

- **shippinginfo() function**—Invokes the *shipping.php* PHP script, passing the *userid* to it. The function also helps in displaying the response generated by *shipping.php*.

- **shipping.php script**—Retrieves existing user information from the *customers* table and displays the necessary shipping input fields.

- **getshipping div**—Displays the *Placing Order* panel shown in Figure 8.6. The response of *shipping.php* script is displayed through this div.

- **saveshipping() function**—Retrieves the shipping information entered by the user and passes it to *saveorder.php,* which saves it to the server-side database tables.

Let's begin by creating *shippinginfo()* function—the code is shown in Listing 8.13.

Listing 8.13. *shippinginfo()* function

```
function shippinginfo(usr)
{
    $('#shipdetails').children().remove();
    $.ajax({
        type:"POST",
        url:"shipping.php",
        data: 'userid='+usr,
        success:function(html){
            $('#shipdetails').append(html);
        }
    });
    jQT.goTo('#getshipping', 'slide');
    return false;
}
```

The sequence of events is as follows:

1. The *userid* of the authenticated user is passed to the *shippinginfo()* function and assigned to the *usr* parameter.

2. Any previous message displayed through *shipdetails* div is removed.

3. The *userid* assigned to the *usr* parameter is passed to *shipping.php,* which retrieves the user's stored information. The script displays the shipping and payment input fields—*shipping address, city, state, country, zipcode, credit card name, number,* and *expiration date.*

4. If the request sent to the PHP script *shipping.php* succeeds, its response is received in the *html* parameter of the *success* callback function.

5. The script response assigned to the *html* parameter is appended to *shipdetails* div for display.

6. The *shipdetails* div is nested inside the *getshipping* div, so the app jumps to the *getshipping* panel with *slide* animation, and displays the *shipping.php* response shown in Figure 8.6(a).

The *shipping.php* script fetches and displays the user information from the *customers* table, and prompts for shipping information. The code of *shipping.php* is shown in Listing 8.14.

Listing 8.14. *shipping.php* script

```php
<?php
    $uid =trim($_REQUEST['userid']);
    $connect=mysql_connect("localhost","root", "mce") or die ("Please check your server
        connection");
    mysql_select_db("shopping");
    $query="Select name, address, city, state, zipcode, emailid, contact_no, country from
        customers where userid ='$uid'";
    $results =mysql_query($query) or die (mysql_query());
    echo '<ul class="rounded">';
    while ($row=mysql_fetch_array($results))
    {
        extract ($row);
        echo '<li>Name: '   . $name . '</li>';
        echo '<li>Address: '   . $address . '</li>';
        echo '<li>City: '   . $city . '</li>';
        echo '<li>State: '   . $state . '</li>';
        echo '<li>Zip Code: '   . $zipcode . '</li>';
        echo '<li>Email Id: '   . $emailid . '</li>';
        echo '<li>Contact Number: '   . $contact_no . '</li>';
    }
    echo '<li><input type="text"  id="shipadd" placeholder="Shipping Address" /></li>';
    echo '<li><input type="text"  id="shipcity" placeholder="Shipping City" /></li>';
    echo '<li><input type="text"  id="shipstate" placeholder="Shipping State" /></li>';
    echo '<li><input type="text"  id="shipcountry" placeholder="Shipping Country" />
        </li>';
    echo '<li><input type="text"  id="shipzipcode" placeholder="Shipping Zip Code" />
        </li>';
    echo '<li><input type="text"  id="cardname" placeholder="Credit Card Name" /></li>';
    echo '<li><input type="text"  id="cardnumber" placeholder="Credit Card Number" />
        </li>';
    echo '<li><input type="text"  id="expirydate" placeholder="Credit Card Expiry Date" />
```

```
        </li>';
    echo '<a class="submit whiteButton" href="#" onclick="saveshipping(\'' . $uid. '\');">
        Place Order</a></li>';
    echo '</ul>';
?>
```

The sequence of events is as follows:

1. The *userid* of the authenticated user is passed to the script through the *shippinginfo()* function.

2. The *userid* is fetched via a *$_REQUEST* array and stored in the *$uid* variable.

3. A connection to the *shopping* database is established.

4. An SQL SELECT statement is executed to fetch the information from the *customers* table matching the supplied *userid* and displayed on the screen through list items.

5. Input text fields, wrapped inside list items, are displayed. These fields allow the user to enter shipping address, shipping city, shipping state, shipping country, shipping zip code, credit card name, credit card number, and credit card expiration date, as shown in Figure 8.6(a).

6. Each input field is given a unique id: *shipadd, shipcity, shipstate, shipcountry, shipzipcode, cardname, cardnumber,* and *expirydate* respectively.

7. The script response is returned to the *shippinginfo()* function (see Listing 8.13) and assigned to the *html* parameter of the *success* callback function.

8. The information received in the *html* parameter is appended to the *shipdetails* div for display.

9. Below the input fields is a *Place Order* button which, when selected, invokes the *saveshipping()* function and passes the *userid* stored in *$uid*.

10. The *saveshipping()* function passes the shipping information to *saveorder.php,* which stores it in the *orders* and *orders_details* server-side database tables.

The *getshipping* div display a panel through which the response of the *shipping.php* PHP script will be displayed. The code of the *getshipping* div is shown in Listing 8.15.

Listing 8.15. *getshipping* div

```
<div id="getshipping">
    <div class="toolbar">
        <h1>Placing Order</h1>
        <a class="cancel" href="#">Cancel</a>
    </div>
    <div id="shipdetails">
    </div>
</div>
```

The *getshipping* div has two divs nested inside it—*toolbar* and *shipdetails*. The *toolbar* div displays a *Cancel* button that cancels the operation and returns to the previous panel. The *toolbar* div also contains a *heading*

1 element that displays the *Placing Order* title in the panel. The *shipdetails* div displays the response generated by *shipping.php*.

When the *Place Order* button is tapped from the *Placing Order* panel, shown in Figure 8.6(b), the *saveshipping* function() is invoked. This function passes the shipping information to *saveorder.php* and is shown in Listing 8.16.

Listing 8.16. *saveshipping()* function

```
function saveshipping(usr)
{
    var shipadd=$('#shipadd').val();
    var shipcity=$('#shipcity').val();
    var shipstate=$('#shipstate').val();
    var shipcountry=$('#shipcountry').val();
    var shipzipcode=$('#shipzipcode').val();
    var cardname=$('#cardname').val();
    var cardnumber=$('#cardnumber').val();
    var expirydate=$('#expirydate').val();
    var data='userid='+usr+'&shipadd='+shipadd+'&shipcity='+shipcity+'&shipstate='+
        shipstate+'&shipcountry='+shipcountry+'&shipzipcode='+shipzipcode+'&cardname='
        +cardname+'&cardnumber='+cardnumber+'&expirydate='+expirydate;
    $('#orderdetails').children().remove();
    $.ajax({
        type:"POST",
        url:"saveorder.php",
        data: data,
        success:function(html){
            $('#orderdetails').append(html);
        }
    });
    jQT.goTo('#thanks', 'slide');
    return false;
}
```

The sequence of events is as follows:

1. The shipping information entered into *shipadd, shipcity, shipstate, shipcountry, shipzipcode, cardname, cardnumber,* and *expiration date* is fetched and stored temporarily in the *shipadd, shipcity, shipstate, shipcountry, shipzipcode, cardname, cardnumber,* and *expirydate* JavaScript variables.

2. The shipping information is then passed to *saveorder.php,* which saves it to the *orders* and *orders_details* server-side databases.

3. The *saveorder.php* script generates a response in the form of a Thanks message and the order number. The order number is displayed to the user for future reference.

4. The response from *saveorder.php* is appended to the *orderdetails* div for display.

5. Because the *orderdetails* div is nested inside the *thanks* div, the app jumps to the *thanks* div with *slide* animation to display the Thanks message and order number generated by *saveorder.php.*

6. The function returns *false* to avoid default browser behavior

Let's finish the chapter by creating the last panel of the application—Thank You.

8.5 Creating the *Thank You* Panel

The *Thank You* panel displays a thanks message and the order number (see Figure 8.7),

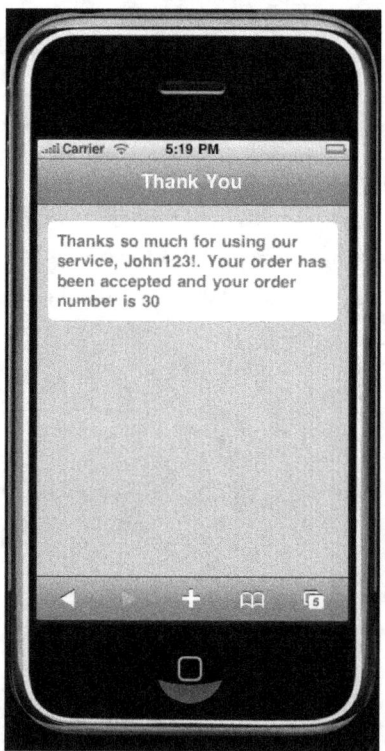

Figure 8.7. *Thank You* **panel displaying a thanks message and order number**

To create the *Thank You* panel and store the shipping information in the server-side database tables, we have to create the following:

saveorder.php script—The script does following:

- Saves the user's shipping information in the *orders* table.

- Generates a Thanks message and order number.

- Searches for all the books in the cart and sends their information individually to *saveorderdetails.php*, which saves them in the *orders_details* table.

152

- Invokes the *makecartempty()* function to empty the cart.

***saveorderdetails.php* script**—Inserts the information of each book in the cart into the *orders_details* table

***makecartempty()* function**—Deletes all the rows from the current user's *shopcart* table.

***thanks* div**—Displays the *Thank You* panel.

Let's start with the *saveorder.php* script, shown in Listing 8.17.

Listing 8.17. *saveorder.php* script

```php
<?php
    session_start();
    $sessid=session_id();
    $today=date("Y-m-d");
    $userid =trim($_REQUEST['userid']);
    $shipadd =trim($_REQUEST['shipadd']);
    $shipcity =trim($_REQUEST['shipcity']);
    $shipstate =trim($_REQUEST['shipstate']);
    $shipcountry =trim($_REQUEST['shipcountry']);
    $shipzipcode =trim($_REQUEST['shipzpcode']);
    $cardname =trim($_REQUEST['cardname']);
    $cardnumber =trim($_REQUEST['cardnumber']);
    $expirydate =trim($_REQUEST['expirydate']);
     $connect=mysql_connect("localhost","root", "mce") or die ("Please check your server
        connection");
    mysql_select_db("shopping");
    $query="INSERT INTO orders (order_date, userid, shipping_address, shipping_city,
        shipping_state, shipping_country, shipping_zipcode, credit_card_name,
        credit_card_number, card_expirydate) VALUES('$today','$userid', '$shipadd',
        '$shipcity', '$shipstate', '$shipcountry', '$shipzipcode', '$cardname',
        '$cardnumber', '$expirydate')";
    $results =mysql_query($query);
    $orderno=mysql_insert_id();
    echo '<ul class="rounded">';
    echo '<li>Thanks so much for using our service, '  . $userid . '!. Your order has been
        accepted and your order number is ' . $orderno . '</li>';
    echo '</ul>';
?>
```

```
<script type="text/javascript">
    var sid="<?php  echo $sessid; ?>";
    var orderno="<?php  echo $orderno; ?>";
    var datab;var shortName = 'tmpCart';
    var version = '1.0';
    var displayName = 'tmpCart';
    var maxSize = 200000;
    datab = openDatabase(shortName, version, displayName, maxSize);
    datab.transaction(
        function(transaction) {
            transaction.executeSql(
                'SELECT cart_sess, cart_isbn,  cart_item_name, cart_qty, cart_price FROM
                    shopcart where cart_sess=?;',[sid],
                function (transaction, result) {
                    for (var i=0; i < result.rows.length; i++) {
                        var row = result.rows.item(i);
                        var data="ordno="+orderno+"&isbn="+row.cart_isbn+"&title="+
                            escape(encodeURI(row.cart_item_name))+"&quantity="+
                            row.cart_qty+ "&price="+row.cart_price;
                        $(document).ready(function(){
                            $.ajax({
                                type:"POST",
                                url:"saveorderdetails.php",
                                data: data,
                            });
                        });
                    }
                }
            );
        }
    );
    makecartempty();
</script>
```

The sequence of events is as follows:

1. The *userid, shipping address, shipping city, shipping state, shipping country, shipping zip code, credit card name, credit card number,* and *expiration date* is passed to *saveorder.php* through the *saveshipping()* function. This information is retrieved using a *$_REQUEST* array and stored in

$userid, $shipadd, $shipcity, $shipstate, $shipcountry, $shipzipcode, $cardname, $cardnumber,* and *$expirydate respectively.

2. The session id of the current user is retrieved and stored in *$sessid* variable.

3. Today's date is extracted and stored in the *$today* variable . This date is the order date.

4. A connection to the server is established and the *shopping* database selected.

5. An SQL INSERT statement is executed to store the shipping information into the *orders* table.

6. Recall from Chapter 1, that the *orders* table has a primary key field, *order_no*, of type integer and is set to auto-increment by 1 on insertion of each row. So, by using the following statement:

   ```
   $orderno = mysql_insert_id();
   ```

   The *order_no* of the *orders table's* newly inserted row is retrieved and stored in the *$orderno* variable.

7. The order number saved in *$orderno* variable will be added to each selected book in the cart while saving them in the *orders_details* table. That is, all the books purchased through an order will have the same order number.

8. A *Thanks* message and order number is returned as the response to *saveshipping()* (see Listing 8.16) and is assigned to the *html* parameter of the *success* callback function.

9. The information assigned to the *html* parameter is then assigned to *orderdetails* div for display.

   Note: We created two tables, *orders* and *orders_details,* for saving the order information. The customer information and order number is stored in the *orders* table. All the books being purchased are stored in the *orders_details* table and will have the same order number.

10. All the cart items (the rows in the *tmpCart*'s *shopcart* table) are passed to the *saveorderdetails.php* script, which saves the information about the order contents.

8.5.1 Passing Non-ASCII Characters

You might have noticed the *escape()* and *encodeURI()* methods used in Listing 8.17 while creating *data* for *saveorderdetails*.php. A book title may contain non-ASCII characters such as an ampersand (&), apostrophe ('), or tilde (~). It's very difficult to pass non-ASCII characters in a query string or URL, because these characters may break the URL. We need to encode these characters using the following methods:

escape() **method**—The *escape()* method is a JavaScript method that returns a string value in Unicode format. It replaces all spaces, punctuations, accented characters, and non-ASCII characters with %xx encoding. where %xx is equivalent to the hexadecimal number representing the character.

encodeURI() **method**—The *encodeURI()* method returns an encoded URI. It replaces all the special characters except : ,/?:@&=+$# with the appropriate UTF escape sequences. We usually need to convert a string to the URI-encoded format to make it suitable for transmission as a query string or part of a URL.

Note: In the target PHP script, *saveorderdetails.php*, we need to use the *urldecode()* method to decode the encoded URI back to the original string

The *saveorderdetails.php* script saves the information about the books in the cart into the *orders_details* table and is shown in Listing 8.18.

Listing 8.18. *saveorderdetails.php* script

```php
<?php
    $ordno =$_REQUEST['ordno'];
    $isbn =trim($_REQUEST['isbn']);
    $title =urldecode(trim($_REQUEST['title']));
    $quantity =$_REQUEST['quantity'];
    $price =$_REQUEST['price'];
    $connect=mysql_connect("localhost","root", "mce") or die ("Please check your server
        connection");
    mysql_select_db("shopping");
    $query="INSERT INTO orders_details (order_no, isbn, title, quantity, price)
        VALUES($ordno,'$isbn', '$title', $quantity, $price)";
    $results =mysql_query($query);
?>
```

The sequence of events is as follows:

- This script retrieves the order number, ISBN, title, quantity, and price sent by *saveorder.php* via a *$_REQUEST* array and saves the data in *$ordno, $isbn, $title, $quantity* and *$price* variables respectively.

- The book title is decoded back to the string using the *urldecode()* method. Remember that the book title was encoded by the *escape()* and *encodeURI()* methods in the *saveorder.php* script before being passed to the *saveorderdetails.php* script.

- A Connection to the MySQL server is established and the *shopping* database is selected.

- An SQL INSERT command is written and executed, which stores the information about the book(s) in the cart into the *orders_details* table.

After saving the data into the *orders_details* server-side database table, there's no point in keeping this information in the *tmpCart* client-side database. Remember that the cart contents data is maintained in the client-side*tmpCart* database's *shopcart* table. So, let's create a *makecartempty()* function that deletes all the rows from *shopcart*. The code of the *makecartempty()* function appears shown in Listing 8.19.

Listing 8.19 *makecartempty()* function

```javascript
function makecartempty()
{
    var sid="<?php  echo $sessid; ?>";
    var datab;var shortName = 'tmpCart';
    var version = '1.0';
    var displayName = 'tmpCart';
    var maxSize = 200000;
    datab = openDatabase(shortName, version, displayName, maxSize);
    datab.transaction(
```

```
function(transaction) {

    transaction.executeSql(

        'delete from shopcart where cart_sess=?;',

        [sid],

        function(){
        },

        displayerrormessage

    );

  }

);

}
```

The sequence of events is as follows:

1. The client-side *tmpCart* database is opened and the connection information is stored in a *datab* object.

2. An SQL DELETE statement is executed via the *transaction's executeSql* method. The statement deletes all the rows from the current user's *shopcart* table.

3. An error message is displayed via the *displayerrormessage* function if the SQL statement fails.

4. The Thanks message and order number are displayed in the *Thank You* panel via the *thanks* div. The code of the *thanks* div is shown in Listing 8.20.

Listing 8.20. *thanks* div

```
<div id="thanks">

    <div class="toolbar">

            <h1>Thank You</h1>

    </div>

    <div id="orderdetails">

    </div>

</div>
```

The *thanks* div has two divs nested inside it—*toolbar* and *orderdetails*. The *toolbar* div contains a *heading 1* element that displays a *Thank You* message. The *orderdetails* div displays the response generated by *saveorder.php*.

8.6 Summary

In this chapter of the book, we learned to write code for the final phase of our sample bookstore web application. The complete code is given in Appendix C.

I've tried my best to keep the code easy to understand in this book and I hope you agree! You now have all the necessary tools for building a small web application store. Have fun creating your own customized programs, and thanks for reading!

Appendix A

Apache Web Server and PHP are automatically installed with the Mac OS X operating system, so you don't have to download them from any other source. All we need to do is activate them.

Apache Web Server

The Apache Web Server can be switched On and Off with two methods:
- From the Terminal window
- From System Preferences

Starting Apache from the Terminal Window

To start or stop the Apache Web Server from the Terminal window, we need to open it and write *sudo apachctl start* at the prompt, as shown in Figure A.1.

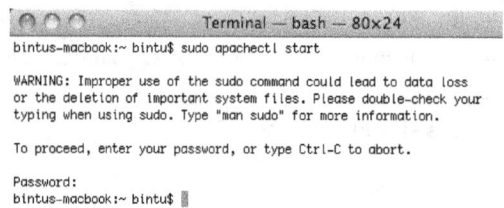

Figure A.1. Starting the Apache Web Server

You will be prompted to enter the administrator's password. If the password is correct, the Apache Web Server will start. You can confirm success by opening the browser and pointing it to the address, *http://localhost*. You should see the output shown in Figure A.2.

It works!

Figure A.2. Output confirming that the Apache Web server is running

Note: The folder where Apache Web server looks for web app files is */Library/WebServer/Documents* folder.

You can also stop the Apache Web Server from the Terminal window with the following command:

```
sudo apachectl stop
```

To confirm that the Apache server is stopped, open the browser and point it again to *http://localhost*. You will see the output shown in Figure A.3.

Note: We will not be asked for an administrator password when stopping the Apache Web Server.

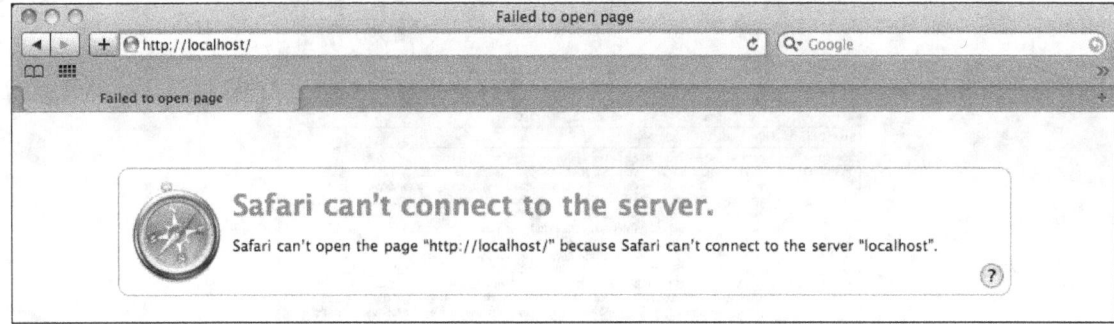

Figure A.3. Output confirming that the Apache Web Server is stopped

Starting Apache from System Preferences

Select *System Preferences* from the Apple menu and click the icon labeled *Sharing*. You'll see a list of services that can be switched on or off. Locate the *Web Sharing* service and check its checkbox to switch it On. You should see a *Web Sharing: On* radio button, which means that the Apache Web Server has started.

To confirm, open the browser and point to the address *http://localhost.* You'll see get the output shown in Figure A.2.

To stop the Apache server from the System Preferences, select the icon labeled *Sharing* and uncheck the *Web Sharing* service . You'll see a *Web Sharing: Off* radio button, which means that the Apache Web Server is now stopped.

Configure PHP

PHP and its popular extensions is included in Leopard. All relevant PHP settings are in the php.ini file. To activate PHP, open the *httpd.conf* file in the */etc/apache2* folder and uncomment the following line:

```
LoadModule php5_module        libexec/apache2/libphp5.so
```

Restart the Apache Web Server by entering the following command:

```
sudo apachectl restart
```

To confirm that PHP is activated, create a file with a *.php* extension and save it in the */Library/WebServer/Documents* folder.

The best way to test that PHP is activated is to create a small file, for example, *a1.php*, containing the following command:

```
<?php phpinfo(); ?>
```

Save the file in the */Library/WebServer/Documents* folder. Then open your browser and point to the *http://localhost/a1.php.*

If the web scripting language PHP is activated, you'll see the output shown in Figure A.4.

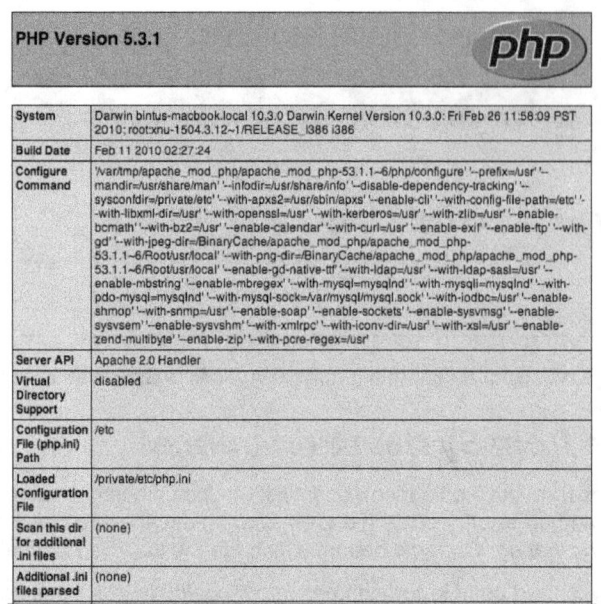

Figure A.4. PHP information displayed via the *phpinfo()* function

Installing MySQL

Open the MySQL downloads page and grab a copy of the Mac OS X 10.6 (x86) installer, then download the *mysql-5.1.48-osx10.6-x86.dmg* disk image file from the Internet. The package contents are shown in Figure A.5.

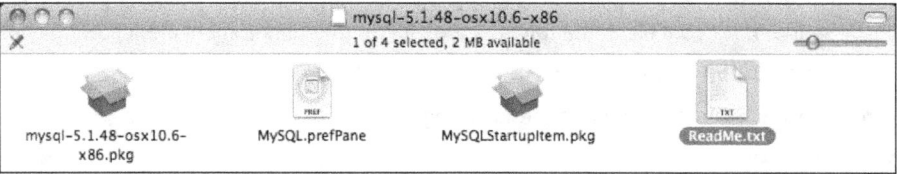

Figure A.5. The *mysql-5.1.48-osx10.6-x86.dmg* package contents

To install the MySQL Package, double-click on the Package icon. This launches the Mac OS X Package Installer, which guides you through the installation.

You can also install a **Preference Pane** that allows MySQL to be stopped and started from *System Preferences*. I suggest you install it, as well. After installation, you'll see its icon in *System Preferences.*

To start MySQL, select the icon. You'll see the screen shown in Figure A.6, which tells you that the MySQL server instance is currently in stopped mode.

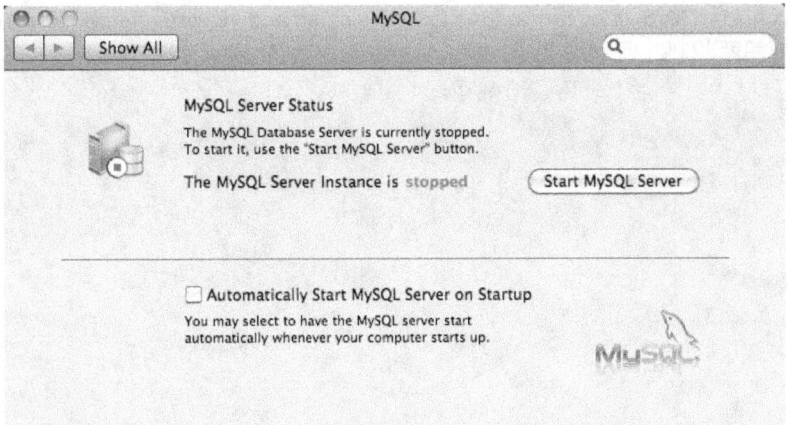

Figure A.6. MySQL in stopped mode

To start a MySQL server instance, select the *Start MySQL Server* button. The MySQL server toggles On, as shown in Figure A.7.

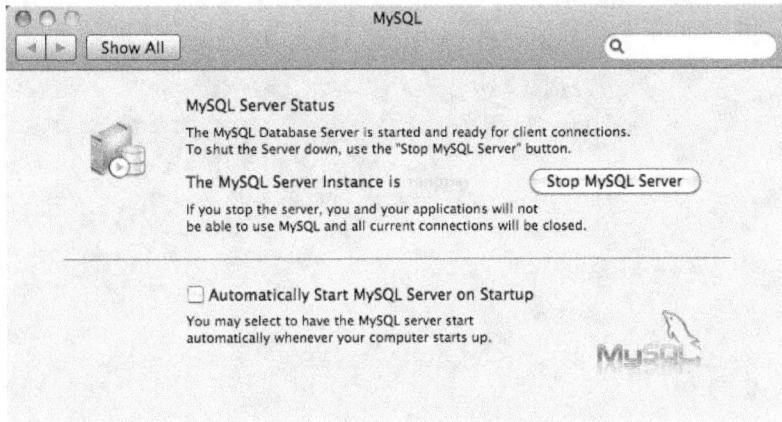

Figure A.7. MySQL in running mode

Starting and Stopping MySQL from Terminal

Use the following command to start MySQL server from the Terminal window:

```
sudo /usr/local/mysql/support-files/mysql.server start
```

To stop MySQL server, use the following command:

```
sudo /usr/local/mysql/support-files/mysql.server stop
```

If you want MySQL to start automatically during system startup, install the *MySQL Startup Item*, available in the disk images file (see Figure A.5). Just double-click the *MySQLStartupItem.pkg* icon and follow the instructions.

> Note: You only have to install the Startup Item once. You won't have to reinstall it if you upgrade the MySQL package later.

The Startup Item for MySQL is installed in the */Library/StartupItems/MySQLCOM* folder. Startup Item installation adds a *MYSQLCOM=-YES-* variable to the */etc/hostconfig* system configuration file. If you want to disable automatic startup of MySQL, simply change this variable to *MYSQLCOM=-NO-*.

After the installation, start MySQL by running the following command in a terminal window. You will need administrator privileges to perform this task.

```
sudo /Library/StartupItems/MySQLCOM/MySQLCOM start
```

B

Appendix B

This chapter covers logging into the MSQL Server and the basic SQL commands you'll need for the sample database.

Logging on to MySQL Server

You first must connect to the MSQL server before you can work with MySQL databases. So switch to the appropriate folder by typing the following command in the Terminal window:

```
cd /usr/local/mysql
```

followed by this command:

```
bin/mysql -h hostname -u username -p
```

- The *hostname* is the name or IP address of the computer on which the MySQL server is running. We can write *-h localhost* or *-h 127.0.0.1* if the MySQL server is running on the current machine.

- The *-u root* argument tells the server that we want to be identified as the MySQL user named *root*. The -u argument is followed by the user name.

- The *-p* argument tells the program to prompt for the password before it tries to connect to the MySQL server.

If the username and password entered are correct, you will be connected to the MySQL server and the MySQL command prompt is displayed. You'll type the executable commands at this prompt. The login procedure is shown in Figure B.1.

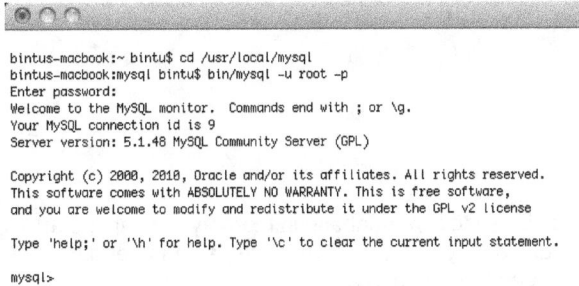

Figure B.1. Logging in to the MySQL server

The MySQL commands we will talk about in this appendix are as follows:

1. Creating databases
2. Showing databases
3. Using databases
4. Showing tables
5. Dropping databases
6. Creating tables

7. Describing table names
8. Dropping tables
9. Inserting table rows
10. Selecting table rows
11. Updating tables
12. Deleting table rows

Create Database

A database is an entity that stores tables, their indexes, foreign key constraints, primary key constraints, and other necessary components. The database houses all the information stored at the back-end of a web application. Having all the parts in one place means you only need to operate on, and back up, a single file instead of multiple tables, indexes, and so on.

Syntax

```
create database database_name;
```

Example

```
mysql> create database shopping;
```

The example creates a database named *shopping*. In place of *shopping*, you can, of course, use any other name you like.

Note: The semi colon (;) is essential after every SQL statement in MSQL.

Show Databases

This command displays all the databases created thus far that currently exist on the MySQL server.

Syntax

```
show databases;
```

Example

```
mysql> show databases;
```

Use Database

This command is used to load the specified database in memory to examine and operate on its components. Only one database can be in use at any one time. When we use another database, the previous database is automatically closed and unloaded from memory.

Syntax

```
use database_name;
```

Example

```
use shopping;
```

When this command is executed and the database loaded, you'll see a *Database changed* message in the Terminal window. In the example, a database named *shopping* is loaded in memory. You can now view, edit or insert information in the database's tables. You can also create new constraints, indexes, and so on.

Show Tables

This command displays all the tables that exist in the currently opened database.

Syntax

```
show tables
```

If the database is empty and has no tables, a message, *Empty set,* is displayed. Otherwise, the table list is shown.

Drop Database

This command deletes the specified database.

Syntax

```
drop database_name
```

Example

```
mysql>drop database shopping;
```

This example deletes the *shopping* database and of its components—tables, indexes, and so on.

Create Table

The first operation usually performed on a database is creating tables. A table is a section of the database that stores information about a person, item entity or object. A table consists of fields or columns. For example, in a table called *products*, the fields or columns might be called *item_code, item_name, description,* and *price.* Each field is defined as a specific data type, which also determines the field size. The field or column in a MySQL table can be defined as any of the data types shown in the table below.

Data Type	Stores
SMALLINT, MEDIUMINT, INT, BIGINT	Integer values
FLOAT	Single-precision floating point values
DOUBLE	Double-precision floating point values
CHAR	Fixed-length strings up to 255 characters
VARCHAR	Variable-length strings up to 255 characters
TINYBLOB, BLOB, MEDIUMBLOB, LONGBLOB	Large blocks of binary data
TINYTEXT, TEXT, MEDIUMTEXT, LONGTEXT	Longer blocks of text data
DATE	Date values
TIME	Time values or durations
DATETIME	Combined date and time values

Syntax

```
mysql>CREATE TABLE table_name (
-> column1 data_type1,
-> column2 data_type2,
-> ...
->);
```

Example

```
mysql> create table product(
    -> id int(6) not null auto_increment,
    -> name varchar(50) not null,
    -> price float,
    -> primary key (id));
Query OK, 0 rows affected (0.20 sec)

mysql> _
```

Figure B.2. Creating a table named *product*

1. The first line says we wish to create a new table named *product*.

2. The second line says that we want a column called *id* that will contain an integer (*int*), and may not be left blank (*not null*). If we don't specify a value for this field when adding a new table row, then MySQL will pick a value that is one more than the highest value in the table so far (*auto_increment*). Finally, this column is to act as a unique identifier for the rows in the table, therefore all the values in this column must be unique for each row.

3. The third line defines a column called *name* that will contain data of character type and its contents may not be left blank.

4. The fourth line defines a column called *price* that will contain a *float* value, that is, a non-integer value.

5. The fifth line confirms that the column *id* specified on the second line is a *primary key,* and hence, it may not be empty.

If the command is given correctly, MySQL creates a table and responds with a *Query OK* message. If you make a typing mistake, MySQL will inform you that there is a problem and you will have to retype the command..

Describe Table Name

This command displays the structure of the specified table. When executed, a list of fields (columns) is displayed, along with their data types. The command shows if a field can store a NULL value; specifies a default value that can be assigned; and which field is set as a primary key.

Syntax

```
describe table_name
```

Example

```
mysql>describe products;
```

The output displays the complete structure of the *products* table, as shown in Figure B.3.

```
mysql> describe product;
+---------+-------------+------+-----+---------+----------------+
| Field   | Type        | Null | Key | Default | Extra          |
+---------+-------------+------+-----+---------+----------------+
| id      | int(6)      | NO   | PRI | NULL    | auto_increment |
| name    | varchar(50) | NO   |     |         |                |
| price   | float       | YES  |     | NULL    |                |
+---------+-------------+------+-----+---------+----------------+
3 rows in set (0.13 sec)
```

Figure B.3 Example of the *Describe* command

Figure B.3 shows that the *product* table contains three columns: *id, name*, and *price*. The primary key is *id* and can neither have a null value nor store a duplicate value. That is, no two rows can have the same *id*. The *name* column can store a name of up to 50 characters. The *price* column can store a decimal because it is declared as a *float* data type.

Drop Table

This command deletes a table in the currently open database. All rows, along with the structure of the table, will be permanently erased.

Syntax

```
mysql>DROP TABLE tablename;
```

Example

```
mysql>drop table product;
```

This command removes the *product* table from the currently active database.

Inserting Table Rows

There are two ways of adding rows to a table. We show both methods below.

```
1) mysql>INSERT INTO table name SET
-> column1 = value1,
-> column2 = value2,
-> ...
->;
```

```
2) mysql>INSERT INTO table name
-> (column, column2, ...)
-> VALUES (value1, value2, ...);
```

Note that in the second form of the INSERT command, the order in which the columns are listed must match the order in which the values are listed. That is, if the sequence of columns in the column list is: *name, price...,* then the VALUES list must be in the same sequence—*name*, *price* and so on. Otherwise, the wrong values will be placed in the columns.

Example :

```
insert into product (name, price) Values ('Camera', 120.99);
```

In this example, the *Camera* value will be stored in the *name* column and the value *120.99* will be stored in the *price* column.

Note : Columns not passed through a VALUES list are set to NULL .

Viewing Table Data

To retrieve rows from a specific table, use the *select SQL*, which has the following syntax:

```
select * [field1, field2...] from table_name [where field1 = expression1 [and/or
field2=expression2 .....]]
```

- If * is used instead of field names, then all the fields/columns of the table are displayed.

- With the help of a *where* clause, we may specify a condition. Only the rows that satisfy the given condition are displayed. If the *where* clause is omitted, then all the rows with the specified fields/columns will be displayed.

- We may specify more than one condition by connecting them with *and* and *or* logical operators.

Examples

This example shows all the rows and columns of the *product* table:

```
select * from product;
```

This example show all the rows of the *product* table containing columns with the name *Camera*:

```
select * from product where name like 'Camera';
```

This example shows the *price* column of the *product* table rows with the name *Camera*:

```
select price from product where name like 'Camera';
```

Modifying Table Contents

The SQL statement *update* is used to modify a table. In this statement, we specify the new value for the column(s), followed by the condition that determines the update range—single, multiple, or all the rows in the table.

Syntax

```
mysql>UPDATE table_name SET
-> column = new_value, ...
->WHERE conditions;
```

Example

```
update product
set price=3700 where name like 'Camera';
```

This example updates to 3700 the *price* column value of all rows in the *product* table containing the name *Camera*.

Note: If the condition specified by the *where* clause is omitted, then all the rows will be updated.

Deleting Table Rows

The *delete* SQL statement deletes a table row.

Syntax

```
mysql>DELETE FROM table_name WHERE conditions;
```

Examples

This example deletes all the rows from the *product* table:

```
delete from product;
```

This example deletes all the rows from the *product* table whose product *name* is Camera.

```
delete from product where name like 'Camera';
```

This example deletes all the rows from the product table whose price=3700.

```
delete from product where price=3700;
```

Quit MySQL

To exit MySQL, type *quit* at the MySQL prompt, followed by the *Enter* key:

```
mysql>quit;
```

MySQL will reply with the message: *Bye and exit.*

Executing SQL Scripts

In this book, I used MySQL Workbench for executing MySQL scripts. Download the latest version from the Internet. The steps for executing SQL script via MySQL Workbench are as follows:

1. Select the *New Connection* option from the Home screen to create a new connection string

2. The dialog box will prompt for the *Connection Name*. Specify the connection name as *shopcartconnection*. *Hostname:*, *Port:*, and *Username:* will be automatically set to *127.0.0.1, 3306* and *root* respectively.

3. Select the *OK* button.

4. You will be prompted to enter the root's password. After entering a password, select the *OK* button.

5. Select the *Edit SQL Script* option from the Home screen.

6. From the *Stored Connection* dropdown list, select the connection name that we just created (*shopcartconnection*) and click the *Continue* button.

7. Select the SQL script file from the list of files on the next screen.

8. Check the *Execute file after opening* checkbox, and click the *Finish* button.

9. You will be prompted to enter root's password. Enter it and click the *OK* button.

10. You will see a dialog box asking for *Line Ending Format*. The *CR* option is selected by default. Select *OK* and the SQL script will be executed.

Appendix C

The sample Book Store web application that we developed in this book consists of 13 files, described in the table below.

File	Usage
index.php	The heart of the application, containing all necessary JavaScript functions and divs.
JavaScript Functions in *index.php*	
showcategories()	Calls the *getcategories.php* script to display book categories
selectedcat()	Calls the *getsubcat.php* script to display subcategories of the selected category
selectedsubcat()	Calls the *getboooks.php* script to display books in the selected category and subcategory
showdetails()	Calls the *getdetails.php* script to display detailed information about the selected book
addtocart()	Inserts the selected book in the client-side *shopcart* table of the *tmpCart* database
displayerrormessage()	Displays error message(s) if the SQL statement fails
updateCart()	Updates the selected book quantity in the client-side *shopcart* table of the *tmpCart* database
deleteCart()	Deletes the selected book, (row) from the client-side *shopcart* table of the *tmpCart* database
dispcart()	Displays the cart contents (rows) in the *shopcart* table of the *tmpCart* database
checkout()	Calls the *checkout.php* script to initiate the checkout procedure
registrationform()	Jumps to the *Create Account* panel to create a new account
createaccount()	Calls the *savecustomer.php* script to save the customer's information in the server-side *customers* table of the *shopping* database
loginform()	Jumps to the Sign In panel, allowing user authentication
checklogin()	Calls the *validatelogin.php* script to compare the user-entered

	userid and *password* with the data stored in the *customers* database table
trylogin()	Removes the error message(s) and allows the user to re-enter a *userid* and *password*
shippinginfo()	Calls the *shipping.php* script, which prompts the user to enter shipping information
saveshipping()	Calls the *saveorder.php* script, which saves the customer's shipping information in the *orders* table and the cart contents in the *orders_details* table
makecartempty()	Deletes all the rows from the client-side *shopcart* table of the *tmpCart* database, thus emptying the cart
readcart()	Counts the quantity of books in the cart and their total price
Divs in index.php	
home	Displays the *Home* panel
bookscategories	Displays the *Categories* panel
subcategories	Displays the *Subcategories* panel
booksdisplay	Displays the *Select Books* panel
bookdetailsdisplay	Displays the *Book Details* panel
showcart	Displays the *Items in Cart* panel
checkout	Displays the *Checking Out* panel
signin	Displays the *Sign In* panel
getshipping	Displays the *Placing Order* panel
thanks	Displays the *Thank You* panel
contactus	Displays the *Contact Us* panel
newarrivals	Displays the *New Arrivals* panel
discountoffers	Displays the *Deep Discount* panel
bestselling	Displays the *Best Selling* panel
giftcards	Displays the *Gift Cards* panel

getcategories.php	Fetches book categories from the *books* table of the *shopping* database
getsubcat.php	Fetches book subcategories belonging to the selected category from the *books* table of the *shopping* database
getbooks.php	Fetches all the books belonging to the selected category and subcategory from the *books* table
getdetails.php	Fetches detailed information of a selected book from the *books* table
cart.php	Maintains the cart
checkout.php	Allows users to sign in or create a new account
createaccount.php	Displays the *Create Account* panel with input fields, allowing user to enter information required to create a new account
savecustomer.php	Saves the customer's information in the *customers* table
validatelogin.php	Checks if the entered *userid* and *password* are correct and displays options to retry or create a new account
shipping.php	Displays input fields to enter shipping information
saveorder.php	Saves shipping information into the *orders* table and generates a new order number
saveorderdetails.php	Saves cart contents into the *orders_details* table

The code of all the files mentioned in the table, as well as the SQL script is provided below.

index.php

```php
<?php
    session_start();
    $sessid=session_id();
?>
<html>
    <head>
        <title>Book Store</title>
        <link type="text/css" rel="stylesheet" media="screen" href="jqtouch/jqtouch.css">
        <link type="text/css" rel="stylesheet" media="screen"
          href="themes/apple/theme.css">
        <script type="text/javascript" src="jqtouch/jquery.1.3.2.min.js"></script>
        <script type="text/javascript" src="jqtouch/jqtouch.js"></script>
        <script type="text/javascript">
            var jQT = new $.jQTouch();
            function showcategories(){
                readcart();
                $('#categories').children().remove();
                $.ajax({
```

```
            type:"POST",
            url:"getcategories.php",
            success:function(html){
                $('#categories').append(html);
                jQT.goTo('#bookscategories', 'slide');
            }
        });
        return false;
    }

    function selectedcat(categ){
        $('#subcat').children().remove();
        $.ajax({
            type:"POST",
            url:"getsubcat.php",
            data: 'category='+categ,
            success:function(html){
                $('#subcat').append(html);
                jQT.goTo('#subcategories', 'slide');
            }
        });
        return false;
    }

    function selectedsubcat(cat, subcat){
        $('#bookslist').children().remove();
        $.ajax({
            type:"POST",
            url:"getbooks.php",
            data: 'category='+cat+'&subcategory='+subcat,
            success:function(html){
                $('#bookslist').append(html);
                jQT.goTo('#booksdisplay', 'slide');
            }
        });
        return false;
    }

    function showdetails(isbn)
    {
        $('#bookdetails').children().remove();
        $.ajax({
            type:"POST",
            url:"getdetails.php",
            data: 'isbn='+isbn,
            success:function(html){
                $('#bookdetails').append(html);
                jQT.goTo('#bookdetailsdisplay', 'slide');
            }
        });
        return false;
    }

    function addtocart(isbn, title, price,qty)
    {
        if(qty >0)
        {
            var sid="<?php  echo $sessid; ?>";
            var datab;var shortName = 'tmpCart';
            var version = '1.0';
            var displayName = 'tmpCart';
            var maxSize = 200000;
            datab = openDatabase(shortName, version, displayName, maxSize);
```

```
datab.transaction(
    function(transaction) {
        transaction.executeSql(
            'CREATE TABLE  IF NOT EXISTS shopcart ' +
                ' (id INTEGER KEY NOT NULL PRIMARY KEY AUTOINCREMENT, ' +
            ' cart_sess varchar(50), cart_isbn varchar(30),
                cart_item_name varchar(100), cart_qty integer,
                cart_price float );'
        );
    }
);
datab.transaction(
    function(transaction) {
        transaction.executeSql(
            'SELECT cart_sess, cart_isbn,  cart_item_name, cart_qty,
                cart_price FROM shopcart where cart_sess=? and
                cart_isbn=?;',[sid, isbn],
                function (transaction, result) {
                    if (result.rows.length >0)
                    {
                        var row = result.rows.item(0);
                        qty=parseInt(qty)+parseInt(row.cart_qty);
                        datab.transaction(
                            function(transaction) {
                                transaction.executeSql(
                                    'update shopcart set cart_qty=? where
                                    cart_sess=? and cart_isbn=?;',
                                    [qty, sid, isbn],
                                    function(){
                                    },
                                    displayerrormessage
                                );
                            }
                        );
                    }
                    else
                    {
                        datab.transaction(
                            function(transaction) {
                                transaction.executeSql(
                                    'INSERT INTO shopcart (cart_sess,
                                    cart_isbn,  cart_item_name, cart_qty,
                                    cart_price) VALUES (?,?,?,?,?);',
                                    [sid, isbn, title, qty, price],
                                    function(){
                                    },
                                    displayerrormessage
                                );
                            }
                        );
                    }
                }
        );
    }
);
}

function displayerrormessage(transaction, error) {
    alert('Error:  '+error.message+' has occurred with Code: '+error.code);
    return true;
}
```

```
function updateCart(isb, qty)
{
    var sid="<?php  echo $sessid; ?>";
    var datab;var shortName = 'tmpCart';
    var version = '1.0';
    var displayName = 'tmpCart';
    var maxSize = 200000;
    datab = openDatabase(shortName, version, displayName, maxSize);
    datab.transaction(
        function(transaction) {
            transaction.executeSql(
                'update shopcart set cart_qty=? where cart_sess=? and
                    cart_isbn=?;',
                [qty, sid, isb],
                function(){
                },
                displayerrormessage
            );
        }
    );
}

function deleteCart(isb)
{
    var sid="<?php  echo $sessid; ?>";
    var datab;var shortName = 'tmpCart';
    var version = '1.0';
    var displayName = 'tmpCart';
    var maxSize = 200000;
    datab = openDatabase(shortName, version, displayName, maxSize);
    datab.transaction(
        function(transaction) {
            transaction.executeSql(
                'delete from shopcart where cart_sess=? and cart_isbn=?;',
                [sid, isb],
                function(){
                },
                displayerrormessage
            );
        }
    );
}

function dispcart()
{
    var total=0;
    var subtot=0;
    var sid="<?php  echo $sessid; ?>";
    var datab;var shortName = 'tmpCart';
    var version = '1.0';
    var displayName = 'tmpCart';
    var maxSize = 200000;
    datab = openDatabase(shortName, version, displayName, maxSize);
    $('#showitems').children().remove();
    datab.transaction(
        function(transaction) {
            transaction.executeSql(
                'SELECT cart_sess, cart_isbn,  cart_item_name, cart_qty,
                cart_price FROM shopcart where cart_sess=?;',[sid],
                function (transaction, result) {
                    if (result.rows.length <=0)
                    {
                        $('#showitems').append('<h1> The Cart is Empty
```

```
                        </h1>');
                    }
                    else
                    {
                        for (var i=0; i < result.rows.length; i++) {
                            subtot=0;
                            var row = result.rows.item(i);
                            $('#showitems').append('<div style="float:left;
                                width:300px;"><label>Book: </label><em>' +
                                row.cart_item_name + '</em></div>');
                            $('#showitems').append('<div style="float:left;
                                width:200px;"> <label style="float:left;" >
                                Quantity: </label><form action="cart.php?isbn='
                                +row.cart_isbn+'&action=update'+'" method=
                                "POST" class="form"><input type="text" name=
                                "quantity" size="6"  value="'+row.cart_qty+'"/>
                                <a class="submit whiteButton" href="#"  style="
                                width: 55px;color:rgba(0,0,0,.9);float:right;
                                "> Update</a> </form><form action="cart.php?
                                isbn='+row.cart_isbn +'&action=delete'+'"
                                method= "POST" class="form"><a style="width:
                                45px;color: rgba(0,0,0,.9); float:left;" href=
                                "#" class="submit whiteButton">Delete </a>
                                </form> </div>');
                            $('#showitems').append('<div style="float:left;
                                width:150px;"> Price: <em>' + row.cart_price +
                                '$</em>');
                            subtot=row.cart_qty*row.cart_price;
                            $('#showitems').append('<div style="width: 150px;
                                float:left; ">Sub Total: <em>' +
                                subtot.toFixed(2) + '$</em><br/><br/>');
                            total=total+subtot;
                        }
                        $('#showitems').append('<div style="width: 150px;
                            float:right;"> Total: <em>' + total.toFixed(2) +
                            '$</em>');
                    }
                },
                displayerrormessage
            );
        }
    );
    return false;
}

function checkout()
{
    $('#checkinfo').children().remove();
    var sid="<?php  echo $sessid; ?>";
    var datab;var shortName = 'tmpCart';
    var version = '1.0';
    var displayName = 'tmpCart';
    var maxSize = 200000;
    datab = openDatabase(shortName, version, displayName, maxSize);
    datab.transaction(
        function(transaction) {
            transaction.executeSql(
                'SELECT cart_sess, cart_isbn,  cart_item_name, cart_qty,
                cart_price FROM shopcart where cart_sess=?;',[sid],
                function (transaction, result) {
                    if(result.rows.length >=1)
                    {
                        $.ajax({
```

```
                        type:"POST",
                        url:"checkout.php",
                        success:function(html){
                            $('#checkinfo').append(html);
                        }
                    });
                }
                else
                {
                    $('#checkinfo').append('<h1> Cart is empty</h1>');
                }
            },
            displayerrormessage
        );
        }
    );
    jQT.goTo('#checkout', 'slide');
    return false;
}

function registrationform()
{
    jQT.goTo('#createacct', 'slide');
}

function createaccount()
{
    var usr=$('#user').val();
    var pswd=$('#passwd').val();
    var cfpswd=$('#confirmpass').val();
    var name=$('#name').val();
    var add=$('#address').val();
    var city=$('#city').val();
    var state=$('#state').val();
    var zip=$('#zipcode').val();
    var email=$('#emailid').val();
    var contact=$('#contactno').val();
    var country=$('#country').val();
    $('#userdetails').children().remove();
    if(usr.length <=0)
    {
        $('#userdetails').append('<p> User id cannot be blank. Please supply
            userid </p>');
        return;
    }
    if(pswd.length <=0)
    {
        $('#userdetails').append('<p> Password cannot be blank. Please supply
            password </p>');
        return;
    }
    if(pswd !=cfpswd)
    {
        $('#userdetails').append('<p> Password and Re-enter password don\'t
            match. Please enter again </p>');
        return;
    }
    if(name.length <=0)
    {
        $('#userdetails').append('<p> Name cannot be blank. Please enter your
            name </p>');
        return;
    }
```

```
      if(contactno.length <=0)
      {
          $('#userdetails').append('<p> Contact Number cannot be blank. Please
              supply contact number </p>');
          return;
      }
      if(emailid.length <=0)
      {
          $('#userdetails').append('<p> Email Address cannot be blank. Please
              supply email address </p>');
          return;
      }
      var data='userid='+usr+'&password='+pswd+'&name='+name+'&add='+add+'&city
          ='+city+'&state='+state+'&zip='+zip+'&email='+email+'&contact='+
          contact+'&country= '+country;
      $('#userdetails').children().remove();
      $.ajax({
          type:"POST",
          url:"savecustomer.php",
          data: data,
          success:function(html){
              $('#userdetails').append(html);
          }
      });
      return false;
}

function loginform()
{
      $('#response').children().remove();
      jQT.goTo('#signin', 'slide');
}

function checklogin()
{
      var usr=$('#userid').val();
      var pswd=$('#password').val();
      $('#response').children().remove();
      if(usr.length <=0)
      {
          $('#response').append('<p> User id cannot be blank. Please supply
              userid </p>');
          return;
      }
      if(pswd.length <=0)
      {
          $('#response').append('<p> Password cannot be blank. Please supply
              password </p>');
          return;
      }
      $('#response').children().remove();
      $.ajax({
          type:"POST",
          url:"validatelogin.php",
          data: 'userid='+usr+'&password='+pswd,
          success:function(html){
              $('#response').append(html);
          }
      });
      return false;
}

function trylogin()
```

```
{
    $('#response').children().remove();
}

function shippinginfo(usr)
{
    $('#shipdetails').children().remove();
    $.ajax({
        type:"POST",
        url:"shipping.php",
        data: 'userid='+usr,
        success:function(html){
            $('#shipdetails').append(html);
        }
    });
    jQT.goTo('#getshipping', 'slide');
    return false;
}

function saveshipping(usr)
{
    var shipadd=$('#shipadd').val();
    var shipcity=$('#shipcity').val();
    var shipstate=$('#shipstate').val();
    var shipcountry=$('#shipcountry').val();
    var shipzipcode=$('#shipzipcode').val();
    var cardname=$('#cardname').val();
    var cardnumber=$('#cardnumber').val();
    var expirydate=$('#expirydate').val();
    var data='userid='+usr+'&shipadd='+shipadd+'&shipcity='+shipcity+
        '&shipstate='+shipstate+'&shipcountry='+shipcountry+'&shipzipcode='+
        shipzipcode+'&cardname='+cardname+'&cardnumber='+cardnumber+
        '&expirydate='+expirydate;
    $('#orderdetails').children().remove();
    $.ajax({
        type:"POST",
        url:"saveorder.php",
        data: data,
        success:function(html){
            $('#orderdetails').append(html);
        }
    });
    jQT.goTo('#thanks', 'slide');
    return false;
}

function makecartempty()
{
    var sid="<?php  echo $sessid; ?>";
    var datab;var shortName = 'tmpCart';
    var version = '1.0';
    var displayName = 'tmpCart';
    var maxSize = 200000;
    datab = openDatabase(shortName, version, displayName, maxSize);
    datab.transaction(
        function(transaction) {
            transaction.executeSql(
                'delete from shopcart where cart_sess=?;',
                [sid],
                function(){
                },
                displayerrormessage
            );
```

179

```
            }
        );
    }

    function readcart()
    {
        var total=0;
        var subtot=0;
        var qtycount=0;
        var sid="<?php  echo $sessid; ?>";
        var datab;var shortName = 'tmpCart';
        var version = '1.0';
        var displayName = 'tmpCart';
        var maxSize = 200000;
        datab = openDatabase(shortName, version, displayName, maxSize);
        $('.totqty').children().remove();
        $('.totprice').children().remove();
        datab.transaction(
            function(transaction) {
                transaction.executeSql(
                    'SELECT cart_sess, cart_isbn,  cart_item_name, cart_qty,
                      cart_price FROM shopcart where cart_sess=?;',[sid],
                    function (transaction, result) {
                        if (result.rows.length <=0)
                        {
                            $('.totqty').append('<p>0 Items</p>');
                            $('.totprice').append('<p>0$</p>');
                        }
                        else
                        {
                            for (var i=0; i < result.rows.length; i++)
                            {
                                subtot=0;
                                var row = result.rows.item(i);
                                qtycount=qtycount+parseInt(row.cart_qty);
                                subtot=row.cart_qty*row.cart_price;
                                total=total+subtot;
                            }
                            $('.totqty').append('<p>' + qtycount +' Items</p>');
                            $('.totprice').append('<p>' + total.toFixed(2) +'
                                $</p>');
                        }
                    },
                    displayerrormessage
                );
            }
        );
    }

    $(function(){
        readcart();
    });
    </script>
</head>
<body>
    <div id="home">
        <h1 style="display: table-cell; width: 190px;background-color:white; color:
            #blue; font: bold 28px Helvetica;"> Book Store </h1>
        <div style="display: table-cell; width: 50px; background-color:white;">
            <img  src="cartfigure.tiff"> </div>
        <a style="display: table-cell; position: absolute; top: 1px; height:50px;
            right:0px; left: 244px; background-color:white;" href="#" onclick=
            "dispcart(); jQT.goTo('#showcart', 'slide');"><div class="totqty"></div>
```

```html
            <div class="totprice"></div></a>
        <div class="toolbar">
            <h1>Home</h1>
        </div>
        <ul class="rounded">
            <li class="arrow"><a href="#" onclick="showcategories();">Books</a></li>
            <li class="arrow"><a href="#contactus">Contact Us</a></li>
            <li class="arrow"><a href="#newarrivals">New Arrivals</a></li>
            <li class="arrow"><a href="#discountoffers">Discount Offers</a></li>
            <li class="arrow"><a href="#bestselling">Best Selling</a></li>
            <li class="arrow"><a href="#giftcards">Gift Cards</a></li>
        </ul>
</div>

<div id="bookscategories">
    <h1 style="display: table-cell; width: 190px;background-color:white; color:
        #blue; font: bold 28px Helvetica; "> Book Store </h1>
    <div style="display: table-cell; width: 50px; background-color:white;">
        <img  src="cartfigure.tiff"> </div>
    <a style="display: table-cell; position: absolute; top: 1px; height:50px;
        right:0px; left: 244px; background-color:white;" href="#" onclick=
        "dispcart(); jQT.goTo('#showcart', 'slide');"><div class="totqty"></div>
        <div class="totprice"></div></a>
    <div class="toolbar">
         <a class="back"  href="#home">Home</a>
        <h1>Categories</h1>
    </div>
    <div id="categories">
    </div>
</div>

<div id="subcategories">
    <h1 style="display: table-cell; width: 190px;background-color:white; color:
        #blue; font: bold 28px Helvetica; "> Book Store </h1>
    <div style="display: table-cell; width: 50px; background-color:white;">
        <img  src="cartfigure.tiff"> </div>
    <a style="display: table-cell; position: absolute; top: 1px; height:50px;
        right:0px; left: 244px; background-color:white;" href="#" onclick=
        "dispcart(); jQT.goTo('#showcart', 'slide');"><div class="totqty">
        </div><div class="totprice"></div></a>
    <div class="toolbar">
            <a class="back"  href="#">Back</a>
        <h1>Subcategories</h1>
    </div>
    <div id="subcat">
    </div>
</div>

<div id="booksdisplay">
    <div class="toolbar">
        <a class="back" href="#">Back</a>
        <h1>Select Books</h1>
    </div>
    <div id="bookslist">
    </div>
</div>

<div id="bookdetailsdisplay">
    <div class="toolbar">
            <a class="back"  href="#">Back</a>
        <h1>Book Details</h1>
    </div>
    <div id="bookdetails">
```

```html
         </div>
      </div>

      <div id="showcart">
         <div class="toolbar">
            <a class="button leftButton" href="#" onclick="showcategories();">
               Shopping</a>
            <a class="button" href="#" onclick="checkout();">Check Out</a>
            <h1>Items in Cart</h1>
         </div>
         <div id="showitems" style="color:black;" >
         </div>
      </div>

      <div id="checkout">
         <div class="toolbar">
            <a  class="cancel" href="#">Cancel</a>
            <h1>Checking Out</h1>
         </div>
         <div id="checkinfo" >
         </div>
      </div>

      <div id="signin">
         <div class="toolbar">
            <a class="button cancel" href="#">Cancel</a>
            <h1>Sign In</h1>
         </div>
         <ul class="rounded">
            <li><input type="text"  id="userid" placeholder="Userid" /></li>
            <li><input type="password"  id="password" placeholder="Password" /></li>
            <a href="#" class="submit whiteButton" onclick="checklogin();">Sign In</a>
         </ul>
         <div id="response">
         </div>
      </div>

      <?php include 'createaccount.php'; ?>

      <div id="getshipping">
         <div class="toolbar">
            <h1>Placing Order</h1>
            <a class="cancel" href="#">Cancel</a>
         </div>
         <div id="shipdetails">
         </div>
      </div>

      <div id="thanks">
         <div class="toolbar">
             <h1>Thank You</h1>
         </div>
         <div id="orderdetails">
         </div>
      </div>

      <div id="contactus">
         <div class="toolbar">
            <h1>Contact Us</h1>
            <a class="back" href="#">Back</a>
         </div>
         <div class="info">
            <p>XYZ Book Company</p>
```

```html
            <p>11 Books Street, NY, NY 10012 </p>
            <p>USA</p>
             Email us: <a href="mailto:bmharwani@yahoo.com"
                target="_blank">bmharwani@yahoo.com</a>
        </div>
    </div>

    <div id="newarrivals">
        <div class="toolbar">
            <h1>New Arrivals</h1>
            <a class="back" href="#">Back</a>
        </div>
        <div class="info">
            <p>We have plenty of new arrivals</p>
            <p>New Books at exciting offers are available now</p>
        </div>
        <ul>
            <li>Linux for Lovers, by Bintu </li>
            <li>Master Unix Shell Programming, by Bintu </li>
            <li>Learn Knitting at Home, by Susan </li>
        </ul>
    </div>

    <div id="discountoffers">
        <div class="toolbar">
            <h1>Deep Discount</h1>
            <a class="back" href="#">Back</a>
        </div>
        <div class="info">
            <p>New Books at the deep discounts are available</p>
            <p>Discount is valid just for few days. So Hurry !!!</p>
        </div>
        <ul>
            <li>Introduction to MS-DOS 1.0, by Bintu </li>
            <li>Buggy Whip Construction, by Bintu </li>
        </ul>
    </div>

    <div id="bestselling">
        <div class="toolbar">
            <h1>Best Selling</h1>
            <a class="back" href="#">Back</a>
        </div>
        <div class="info">
            <p>Following is the list of the best selling books</p>
            <p>These books range from Computers to Story & Fiction</p>
        </div>
    </div>

    <div id="giftcards">
        <div class="toolbar">
            <h1>Gift Cards</h1>
            <a class="back" href="#">Back</a>
        </div>
        <div class="info">
            <p>Attractive Festive Gift Cards available at attractive prices</p>
            <p>Free shipping offers for few days</p>
        </div>
    </div>
  </body>
</html>
```

getcategories.php

```php
<?php
   $connect=mysql_connect("localhost","root", "mce") or die ("Please check your server
      connection");
   mysql_select_db("shopping");
   $query="Select distinct category from books";
   $results =mysql_query($query) or die (mysql_query());
   if(mysql_num_rows($results)==0)
   {
      echo '
         <ul>
            <li>No books found</li>
         </ul>';
   }
   else
   {
      echo '<ul class="rounded">';
      while ($row=mysql_fetch_array($results))
      {
         extract ($row);
         echo '
            <li class="arrow"><a href="#" onclick="javascript:selectedcat(\'' .
               urlencode($category) .'\');"> ' . $category . '</a></li>';
      }
      echo '</ul>';
   }
?>
```

getsubcat.php

```php
<?php
   $category=$_POST['category'];
    $connect=mysql_connect("localhost","root", "mce") or die ("Please check your server
       connection");
   mysql_select_db("shopping");
   $query="Select distinct subcategory from books where category ='$category'";
   $results =mysql_query($query) or die (mysql_error());
   if($results)
   {
      echo '<ul class="rounded">';
      while ($row=mysql_fetch_array($results))
      {
         extract ($row);
         echo '<li class="arrow">';
         echo "<a href=\"#\" onclick=\"javascript:selectedsubcat('" .
            urlencode($category). "','". urlencode($subcategory) ."');\">" .
            $subcategory . "</a></li>";
      }
      echo "</ul>";
   }
   else
   {
      echo '<ul class="rounded">';
      echo "<li> No Subcategories found </li>";
      echo "</ul>";
   }
?>
```

getbooks.php

```php
<?php
```

```php
    $cat =trim($_REQUEST['category']);
    $subcat =trim($_REQUEST['subcategory']);
    $connect=mysql_connect("localhost","root", "mce") or die ("Please check your server
        connection");
    mysql_select_db("shopping");
    $query="Select isbn, title, author1, author2, author3, price, image from books where
        category =\"$cat\" and subcategory = \"$subcat\"";
    $results =mysql_query($query) or die (mysql_query());
    if(mysql_num_rows($results)>0)
    {
        while ($row=mysql_fetch_array($results))
        {
            extract ($row);
            echo '<fieldset style="background-color:white; color:black;">';
            echo '<form action="cart.php?isbn=' . $isbn . '&title=' . urlencode($title) .
                '&price=' . $price .'&action=add' . '" method="POST" class="form">';
            echo '<img src=' . $image .'>';
            echo '<h3>' . $title . '</h3>';
            echo '<h4>' . $author1 . '</h4>';
            echo '<label>Price: </label>';
            echo '<em>' . $price . '</em><br/>';
            echo '<label>Quantity:   </label><input type="text" name="quantity" value="1"
                style="height:22px;" size="6"/>';
            echo '<a class="whiteButton" href="#" onclick="showdetails(\'' . $isbn .
                '\');"> Show Details</a>';
            echo '<a class="submit whiteButton" href="#"> Add To Cart</a>';
            echo '</form>';
            echo '</fieldset>';
        }
    }
    else
    {
        echo '<ul class="rounded">';
        echo "<li> No Books found in this Subcategory</li>";
        echo "</ul>";
    }
?>
```

getdetails.php

```php
<?php
    $isbn =trim($_REQUEST['isbn']);
    $connect=mysql_connect("localhost","root", "mce") or die ("Please check your server
        connection");
    mysql_select_db("shopping");
    $query="Select isbn, title, author1, author2, author3, publisher,
        publish_date_edition, price, image, description from books where isbn ='$isbn'";
    $results =mysql_query($query) or die (mysql_query());
    while ($row=mysql_fetch_array($results))
    {
        extract ($row);
        echo '<fieldset style="background-color:white; color:black;">';
        echo '<form action="cart.php?isbn=' . $isbn . '&title=' . urlencode($title) .
            '&price=' . $price . '&action=add' . '" method="POST">';
        echo '<img src=' . $image .'>';
        echo '<h3>' . $title . ' by </h3>';
        echo '<h4>' . $author1 . '</h4>';
        if($author2 !='NULL')
            echo '<h4>' . $author2 . '</h4>';
        if($author3 !='NULL')
            echo '<h4>' . $author3 . '</h4>';
        echo '<label>Publisher :</label><h4>' . $publisher . '</h4>';
        echo '<h4>' . $publish_date_edition . '</h4>';
```

```php
        echo '<label>Price: </label>';
        echo '<em>' . $price . '</em><br/>';
        echo '<label>Book Details :</label><h4>' . $description . '</h4>';
         echo '<label>Quantity :</label><input type="text" style="height:22px;"
            name="quantity" value="1" />';
        echo '<a class="submit whiteButton" href="#" onclick="this.form.submit();"> Add To
            Cart</a>';
        echo '</form>';
        echo '</fieldset>';}
?>
```

cart.php

```php
<?php
    $isbn = trim($_REQUEST['isbn']);
    $title=$_REQUEST['title'];
    $qty=$_REQUEST['quantity'];
    $price=$_REQUEST['price'];
    $action=trim($_REQUEST['action']);
?>
<div id="cartupdated">
    <h1 style="display: table-cell; width: 190px;background-color:white; color: #blue;
        font: bold 28px Helvetica;"> Book Store </h1>
    <a style="display: table-cell; width: 150px; background-color:white;" href="#"
        onclick="dispcart(); jQT.goTo('#showcart', 'slide');"><img  src="cartfigure.tiff">
        Show Cart </a>
    <div class="toolbar">
        <h1>Cart Updated</h1>
        <a class="button leftButton" href="#" onclick="showcategories();"> Shopping</a>
        <a class="button" href="#" onclick="checkout();">Check Out</a>
    </div>
</div>
<script type="text/javascript">
    var sbn="<?php  echo $isbn; ?>";
    var tit="<?php echo $title; ?>";
    var qt="<?php echo $qty; ?>";
    var pr="<?php echo $price; ?>";
    var action="<?php echo $action; ?>";
    if(action=="add") addtocart(sbn,tit,pr,qt);
    if(action=="update")
    {
        if(qt>0) updateCart(sbn, qt);
        else deleteCart(sbn);
    }
    if(action=="delete") deleteCart(sbn);
</script>
```

checkout.php

```php
<?php
    $uid=$_SESSION['userid'];
    $pwd=$_SESSION['password'];
    echo '<ul class="rounded">';
    if ((isset($_SESSION['userid']) && $_SESSION['userid'] != "") ||
        (isset($_SESSION['password']) && $_SESSION['password'] != ""))
    {
        echo'<li>If you are over with Shopping, please provide shipping information</li>';
        echo '<li><a href="#" class="submit whiteButton" onclick="shippinginfo(\'' . $uid
            . '\');">Supply Shipping Info</a></li>';
    }
    else
    {
```

```
    echo'<li>You are not Signed in yet</li>';
     echo '<li><a href="#" class="submit whiteButton" onclick="loginform();">Sign In
        </a></li>';
    echo '<li><a href="#" class="submit whiteButton" onclick="registrationform();">
        Create Account</a></li>';
  }
  echo '</ul>';
?>
```

createaccount.php

```
<div id="createacct">
   <div class="toolbar">
      <h1>Create Account</h1>
      <a class="button cancel" href="#">Cancel</a>
   </div>
   <ul class="rounded">
      <li><input type="text" id="user" placeholder="userid" /></li>
      <li><input type="password" id="passwd" placeholder="password" /></li>
      <li><input type="password" id="confirmpass" placeholder="Re-enter password"/>
      </li>
      <li><input type="text"  id="name" placeholder="Name" /></li>
      <li><input type="text"  id="address" placeholder="Address" /></li>
      <li><input type="text"  id="city" placeholder="City" /></li>
      <li><input type="text"  id="state" placeholder="State" /></li>
      <li><input type="text"  id="zipcode" placeholder="Zip Code" /></li>
      <li><input type="text"  id="emailid" placeholder="Email Id" /></li>
      <li><input type="text" id="contactno" placeholder="Contact No" /></li>
      <li><input type="text" id="country" placeholder="Country" /></li>
      <a href="#" class="submit whiteButton" onclick="createaccount();">Submit</a>
   </ul>
   <div id="userdetails" >
   </div>
</div>
```

savecustomer.php

```
<?php
   $uid =trim($_REQUEST['userid']);
   $pswd =trim($_REQUEST['password']);
   $name =trim($_REQUEST['name']);
   $add =trim($_REQUEST['add']);
   $city =trim($_REQUEST['city']);
   $state =trim($_REQUEST['state']);
   $zip =trim($_REQUEST['zip']);
   $email =trim($_REQUEST['email']);
   $contact =trim($_REQUEST['contact']);
   $country =trim($_REQUEST['country']);
   $connect=mysql_connect("localhost","root", "mce") or die ("Please check your server
      connection");
   mysql_select_db("shopping");
   $query="INSERT INTO customers (userid, password, name, address, city, state, zipcode,
      emailid, contact_no, country) VALUES('$uid', '$pswd', '$name', '$add', '$city',
      '$state', '$zip', '$email','$contact', '$country')";
   $results =mysql_query($query);
   echo '<p>Congratulations '  . $uid . '!. Your account is created </p>';
   echo '<p>Select the below button to Sign In </p>';
   echo '<a href="#" class="submit whiteButton" onclick="loginform();">Sign In</a>';
?>
```

validatelogin.php

```php
<?php
    $uid =trim($_REQUEST['userid']);
    $pswd =trim($_REQUEST['password']);
    $connect=mysql_connect("localhost","root", "mce") or die ("Please check your server
        connection");
    mysql_select_db("shopping");
    $query="Select userid, password from customers where userid ='$uid' and password =
        '$pswd'";
    $results =mysql_query($query) or die (mysql_query());
    if(mysql_num_rows($results)>0)
    {
        echo 'Welcome ' . $uid . '!!';
        echo '<a href="#" class="submit whiteButton" onclick="shippinginfo(\'' . $uid .
            '\');">Supply Shipping Info</a>';
    }
    else
    {
        echo '<p>Sorry the userid or password is incorrect</p>';
        echo '<a href="#" class="submit whiteButton" onclick="trylogin();">Try Again</a>';
         echo '<a href="#" class="submit whiteButton" onclick="registrationform();">Create
            Account</a>';
    }
?>
```

shipping.php

```php
<?php
    $uid =trim($_REQUEST['userid']);
    $connect=mysql_connect("localhost","root", "mce") or die ("Please check your server
        connection");
    mysql_select_db("shopping");
    $query="Select name, address, city, state, zipcode, emailid, contact_no, country from
        customers where userid ='$uid'";
    $results =mysql_query($query) or die (mysql_query());
    echo '<ul class="rounded">';
    while ($row=mysql_fetch_array($results))
    {
        extract ($row);
        echo '<li>Name: '   . $name . '</li>';
        echo '<li>Address: '   . $address . '</li>';
        echo '<li>City: '   . $city . '</li>';
        echo '<li>State: '   . $state . '</li>';
        echo '<li>Zip Code: '   . $zipcode . '</li>';
        echo '<li>Email Id: '   . $emailid . '</li>';
        echo '<li>Contact Number: '   . $contact_no . '</li>';
    }
    echo '<li><input type="text"  id="shipadd" placeholder="Shipping Address" /></li>';
    echo '<li><input type="text"  id="shipcity" placeholder="Shipping City" /></li>';
    echo '<li><input type="text"  id="shipstate" placeholder="Shipping State" /></li>';
     echo '<li><input type="text"  id="shipcountry" placeholder="Shipping Country"/></li>';
    echo '<li><input type="text"  id="shipzipcode" placeholder="Shipping Zip Code"/>
        </li>';
    echo '<li><input type="text"  id="cardname" placeholder="Credit Card Name" /></li>';
    echo '<li><input type="text"  id="cardnumber" placeholder="Credit Card Number"/>
        </li>';
    echo '<li><input type="text"  id="expirydate" placeholder="Credit Card Expiry Date"/>
        </li>';
    echo '<a class="submit whiteButton" href="#" onclick="saveshipping(\'' . $uid. '\');">
        Place Order</a></li>';
    echo '</ul>';
?>
```

saveorder.php

```php
<?php
    session_start();
    $sessid=session_id();
    $today=date("Y-m-d");
    $userid =trim($_REQUEST['userid']);
    $shipadd =trim($_REQUEST['shipadd']);
    $shipcity =trim($_REQUEST['shipcity']);
    $shipstate =trim($_REQUEST['shipstate']);
    $shipcountry =trim($_REQUEST['shipcountry']);
    $shipzipcode =trim($_REQUEST['shipzpcode']);
    $cardname =trim($_REQUEST['cardname']);
    $cardnumber =trim($_REQUEST['cardnumber']);
    $expirydate =trim($_REQUEST['expirydate']);
    $connect=mysql_connect("localhost","root", "mce") or die ("Please check your server
        connection");
    mysql_select_db("shopping");
    $query="INSERT INTO orders (order_date, userid, shipping_address, shipping_city,
        shipping_state, shipping_country, shipping_zipcode, credit_card_name,
        credit_card_number, card_expirydate) VALUES('$today','$userid', '$shipadd',
        '$shipcity', '$shipstate', '$shipcountry', '$shipzipcode', '$cardname',
        '$cardnumber', '$expirydate')";
    $results =mysql_query($query);
    $orderno=mysql_insert_id();
    echo '<ul class="rounded">';
    echo '<li>Thanks so much for using our service, '  . $userid . '!. Your order has been
        accepted and your order number is ' . $orderno . '</li>';
    echo '</ul>';
?>
<script type="text/javascript">
    var sid="<?php  echo $sessid; ?>";
    var orderno="<?php  echo $orderno; ?>";
    var datab;var shortName = 'tmpCart';
    var version = '1.0';
    var displayName = 'tmpCart';
    var maxSize = 200000;
    datab = openDatabase(shortName, version, displayName, maxSize);
    datab.transaction(
        function(transaction) {
            transaction.executeSql(
                'SELECT cart_sess, cart_isbn,  cart_item_name, cart_qty, cart_price FROM
                    shopcart where cart_sess=?;',[sid],
                function (transaction, result) {
                    for (var i=0; i < result.rows.length; i++) {
                        var row = result.rows.item(i);
                        var data="ordno="+orderno+"&isbn="+row.cart_isbn+"&title="
                            +escape(encodeURI(row.cart_item_name))+"&quantity="+
                            row.cart_qty+"&price="+row.cart_price;
                        $.ajax({
                            type:"POST",
                            url:"saveorderdetails.php",
                            data: data,
                        });
                    }
                }
            );
        }
    );
    makecartempty();
</script>
```

saveorderdetails.php

```php
<?php
    $ordno =$_REQUEST['ordno'];
    $isbn =trim($_REQUEST['isbn']);
    $title =urldecode(trim($_REQUEST['title']));
    $quantity =$_REQUEST['quantity'];
    $price =$_REQUEST['price'];
    $connect=mysql_connect("localhost","root", "mce") or die ("Please check your server
        connection");
    mysql_select_db("shopping");
    $query="INSERT INTO orders_details (order_no, isbn, title, quantity, price)
        VALUES($ordno,'$isbn', '$title', $quantity, $price)";
    $results =mysql_query($query);
?>
```

SQL Script for creating the database, tables, and dummy record insertion

```sql
create database shopping;
use shopping;

create table books (
isbn varchar(30) not null,
title varchar(100) not null,
author1 varchar(50) not null,
author2 varchar(50),
author3 varchar(50),
category varchar(50),
subcategory varchar(50),
quantity smallint,
publisher varchar(100),
publish_date_edition  varchar(50),
price float,
image varchar(50),
description text,
primary key(isbn));

create table customers (
userid varchar(50) not null,
password varchar(50),
name varchar(50),
address varchar(200),
city varchar(50),
state varchar(50),
zipcode varchar(12),
emailid varchar(50),
contact_no varchar(50),
country varchar(50),
primary key(userid));

create table orders (
order_no integer not null auto_increment,
order_date date,
userid varchar(50),
shipping_address varchar(200),
shipping_city varchar(50),
shipping_state varchar(50),
shipping_country varchar(50),
shipping_zipcode varchar(15),
credit_card_name varchar(30),
credit_card_number varchar(15),
card_expirydate date,
primary key (order_no));
```

```
create table orders_details (
order_no integer not null,
isbn varchar(30) not null,
title varchar(100) not null,
quantity integer not null,
price float
);

insert into books (isbn, title, author1, author2, author3, category, subcategory,
quantity, publisher, publish_date_edition, price, image, description) values('111-1-1111-
1111-0', 'Red Queen', 'B.M.Harwani','NULL', 'NULL','Literature & Fiction', 'Drama', 100,
'Microchip Education','January 2010 : First Edition', 15.99,'images/111-1-1111-1111-
0.jpg','This collection of short stories, including many new translations, is the first
to span the whole of Japan modern era from the end of the nineteenth century to the
present day');

insert into books (isbn, title, author1, author2, author3, category, subcategory,
quantity,  publisher, publish_date_edition, price, image, description) values('111-1-
1111-1111-1', 'Last Block', 'B.M.Harwani','NULL', 'NULL','Literature & Fiction', 'Drama',
100, 'Microchip Education','Feburary 2010 : First Edition', 19.99,'images/111-1-1111-
1111-1.jpg','This collection of short stories, including many new translations');

insert into books (isbn, title, author1, author2, author3, category, subcategory,
quantity,  publisher, publish_date_edition, price, image, description) values('111-1-
1111-1111-2', 'All New Tales', 'B.M.Harwani','NULL', 'NULL','Literature & Fiction',
'Drama', 100, 'Microchip Education','April 2010 : First Edition', 39.99,'images/111-1-
1111-1111-2.jpg','This collection of short stories, including many new translations');

insert into books (isbn, title, author1, author2, author3, category, subcategory,
quantity,  publisher, publish_date_edition, price, image, description) values('111-1-
1111-1111-3', 'The Fixer', 'B.M.Harwani','NULL', 'NULL','Literature & Fiction', 'Essays',
100, 'Microchip Education','March 2010 : First Edition', 25.99,'images/111-1-1111-1111-
3.jpg','This collection of short stories, including many new translations');

insert into books (isbn, title, author1, author2, author3, category, subcategory,
quantity,  publisher, publish_date_edition, price, image, description) values('111-1-
1111-1111-4', 'Girl with Tattoo', 'B.M.Harwani','Bintu Mott', 'NULL','Literature &
Fiction', 'Essays', 100, 'Microchip Education','May 2010 : First Edition',
20.99,'images/111-1-1111-1111-4.jpg','This collection of short stories, including many
new translations');

insert into books (isbn, title, author1, author2, author3, category, subcategory,
quantity,  publisher, publish_date_edition, price, image, description) values('111-1-
1111-1111-5', 'Girl Who Kicked', 'B.M.Harwani','Bintu Mott', 'David Mortiz','Literature &
Fiction', 'Letters & Correspondence', 100, 'Microchip Education','June 2010 : First
Edition', 35.99,'images/111-1-1111-1111-5.jpg','This collection of short stories,
including many new translations');

insert into books (isbn, title, author1, author2, author3, category, subcategory,
quantity,  publisher, publish_date_edition, price, image, description) values('111-1-
1111-1111-6', 'Boy Who Set Fire', 'B.M.Harwani','NULL', 'NULL','Literature & Fiction',
'Poetry', 100, 'Microchip Education','March 2010 : First Edition', 25.99,'images/111-1-
1111-1111-6.jpg','This collection of short stories, including many new translations');

insert into books (isbn, title, author1, author2, author3, category, subcategory,
quantity,  publisher, publish_date_edition, price, image, description) values('111-1-
1111-1111-7', 'The Dawn', 'B.M.Harwani','Bintu Mott', 'NULL','Literature & Fiction',
'Women\'s Fiction', 100, 'Microchip Education','May 2010 : First Edition',
20.99,'images/111-1-1111-1111-7.jpg','This collection of short stories, including many
new translations');

insert into books (isbn, title, author1, author2, author3, category, subcategory,
quantity,  publisher, publish_date_edition, price, image, description) values('111-1-
```

1111-1111-8', 'Two Sisters', 'B.M.Harwani','Bintu Mott', 'David Mortiz','Literature & Fiction', 'World Literature', 100, 'Microchip Education','June 2010 : First Edition', 35.99,'images/111-1-1111-1111-8.jpg','This collection of short stories, including many new translations');

insert into books (isbn, title, author1, author2, author3, category, subcategory, quantity, publisher, publish_date_edition, price, image, description) values('111-1-1111-1120-0', 'Perfect Plants', 'B.M.Harwani','NULL', 'NULL','Home & Garden', 'Crafts & Hobbies', 100, 'Microchip Education','January 2010 : First Edition', 15.99,'images/111-1-1111-1120-0.jpg','Bigger and better with more growing information, plant profiles, and photos');

insert into books (isbn, title, author1, author2, author3, category, subcategory, quantity, publisher, publish_date_edition, price, image, description) values('111-1-1111-1121-0', 'How to Grow in Small Space', 'B.M.Harwani','NULL', 'NULL','Home & Garden', 'Crafts & Hobbies', 100, 'Microchip Education','Feburary 2010 : First Edition', 19.99,'images/111-1-1111-1121-0.jpg','Bigger and better with more growing information, plant profiles, and photos');

insert into books (isbn, title, author1, author2, author3, category, subcategory, quantity, publisher, publish_date_edition, price, image, description) values('111-1-1111-1122-0', 'Lifetime Gardening', 'B.M.Harwani','NULL', 'NULL','Home & Garden', 'Antiques & Collectibles', 100, 'Microchip Education','April 2010 : First Edition', 39.99,'images/111-1-1111-1122-0.jpg','Bigger and better with more growing information, plant profiles, and photos');

insert into books (isbn, title, author1, author2, author3, category, subcategory, quantity, publisher, publish_date_edition, price, image, description) values('111-1-1111-1123-0', 'Home keeping A Must Have', 'B.M.Harwani','NULL', 'NULL','Home & Garden', 'Antiques & Collectibles', 100, 'Microchip Education','March 2010 : First Edition', 25.99,'images/111-1-1111-1123-0.jpg','Bigger and better with more growing information, plant profiles, and photos');

insert into books (isbn, title, author1, author2, author3, category, subcategory, quantity, publisher, publish_date_edition, price, image, description) values('111-1-1111-1124-0', 'Designing A Home', 'B.M.Harwani','Bintu Mott', 'NULL','Home & Garden', 'Interior Design', 100, 'Microchip Education','May 2010 : First Edition', 20.99,'images/111-1-1111-1124-0.jpg','Bigger and better with more growing information, plant profiles, and photos');

insert into books (isbn, title, author1, author2, author3, category, subcategory, quantity, publisher, publish_date_edition, price, image, description) values('111-1-1111-1125-0', 'Gardening A Complete Guide', 'B.M.Harwani','Bintu Mott', 'David Mortiz','Home & Garden', 'Interior Design', 100, 'Microchip Education','June 2010 : First Edition', 35.99,'images/111-1-1111-1125-0.jpg','Bigger and better with more growing information, plant profiles, and photos');

insert into books (isbn, title, author1, author2, author3, category, subcategory, quantity, publisher, publish_date_edition, price, image, description) values('111-1-1111-1126-0', 'Growing Herbs At Home', 'B.M.Harwani','NULL', 'NULL','Home & Garden', 'Home Design', 100, 'Microchip Education','March 2010 : First Edition', 25.99,'images/111-1-1111-1126-0.jpg','Bigger and better with more growing information, plant profiles, and photos');

insert into books (isbn, title, author1, author2, author3, category, subcategory, quantity, publisher, publish_date_edition, price, image, description) values('111-1-1111-1127-0', 'Master Craftsman', 'B.M.Harwani','Bintu Mott', 'NULL','Home & Garden', 'Home Design', 100, 'Microchip Education','May 2010 : First Edition', 20.99,'images/111-1-1111-1127-0.jpg','Bigger and better with more growing information, plant profiles, and photos');

insert into books (isbn, title, author1, author2, author3, category, subcategory, quantity, publisher, publish_date_edition, price, image, description) values('111-1-

1111-1131-0', 'Computer & Common Sense', 'B.M.Harwani','NULL', 'NULL','Computers & Internet', 'General', 100, 'Microchip Education','January 2010 : First Edition', 15.99,'images/111-1-1111-1131-0.jpg','jQuery has been around for a few years now and is already pretty much the #1 or #2 JavaScript library used in websites today');

insert into books (isbn, title, author1, author2, author3, category, subcategory, quantity, publisher, publish_date_edition, price, image, description) values('111-1-1111-1131-1', 'Introduction to Computers', 'B.M.Harwani','NULL', 'NULL','Computers & Internet', 'General', 100, 'Microchip Education','Feburary 2010 : First Edition', 19.99,'images/111-1-1111-1131-1.jpg','jQuery has been around for a few years now and is already pretty much the #1 or #2 JavaScript library used in websites today');

insert into books (isbn, title, author1, author2, author3, category, subcategory, quantity, publisher, publish_date_edition, price, image, description) values('111-1-1111-1131-2', 'Computer for Beginnners', 'B.M.Harwani','NULL', 'NULL','Computers & Internet', 'General', 100, 'Microchip Education','April 2010 : First Edition', 39.99,'images/111-1-1111-1131-2.jpg','jQuery has been around for a few years now and is already pretty much the #1 or #2 JavaScript library used in websites today');

insert into books (isbn, title, author1, author2, author3, category, subcategory, quantity, publisher, publish_date_edition, price, image, description) values('111-1-1111-1131-3', 'Visual Basic Programming', 'B.M.Harwani','NULL', 'NULL','Computers & Internet', 'Programming', 100, 'Microchip Education','March 2010 : First Edition', 25.99,'images/111-1-1111-1131-3.jpg','jQuery has been around for a few years now and is already pretty much the #1 or #2 JavaScript library used in websites today');

insert into books (isbn, title, author1, author2, author3, category, subcategory, quantity, publisher, publish_date_edition, price, image, description) values('111-1-1111-1131-4', 'C# Programming', 'B.M.Harwani','Bintu Mott', 'NULL','Computers & Internet', 'Programming', 100, 'Microchip Education','May 2010 : First Edition', 20.99,'images/111-1-1111-1131-4.jpg','jQuery has been around for a few years now and is already pretty much the #1 or #2 JavaScript library used in websites today');

insert into books (isbn, title, author1, author2, author3, category, subcategory, quantity, publisher, publish_date_edition, price, image, description) values('111-1-1111-1131-5', 'C++ for Beginners', 'B.M.Harwani','Bintu Mott', 'David Mortiz','Computers & Internet', 'Programming', 100, 'Microchip Education','June 2010 : First Edition', 35.99,'images/111-1-1111-1131-5.jpg','jQuery has been around for a few years now and is already pretty much the #1 or #2 JavaScript library used in websites today');

insert into books (isbn, title, author1, author2, author3, category, subcategory, quantity, publisher, publish_date_edition, price, image, description) values('111-1-1111-1131-6', 'Programming in C ', 'B.M.Harwani','NULL', 'NULL','Computers & Internet', 'Programming', 100, 'Microchip Education','March 2010 : First Edition', 25.99,'images/111-1-1111-1131-6.jpg','jQuery has been around for a few years now and is already pretty much the #1 or #2 JavaScript library used in websites today');

insert into books (isbn, title, author1, author2, author3, category, subcategory, quantity, publisher, publish_date_edition, price, image, description) values('111-1-1111-1131-7', 'Mastering SQL Server', 'B.M.Harwani','Bintu Mott', 'NULL','Computers & Internet', 'Databases', 100, 'Microchip Education','May 2010 : First Edition', 20.99,'images/111-1-1111-1131-7.jpg','jQuery has been around for a few years now and is already pretty much the #1 or #2 JavaScript library used in websites today');

insert into books (isbn, title, author1, author2, author3, category, subcategory, quantity, publisher, publish_date_edition, price, image, description) values('111-1-1111-1131-8', 'Database Handling in Oracle', 'B.M.Harwani','Bintu Mott', 'David Mortiz','Computers & Internet', 'Databases', 100, 'Microchip Education','June 2010 : First Edition', 35.99,'images/111-1-1111-1131-8.jpg','jQuery has been around for a few years now and is already pretty much the #1 or #2 JavaScript library used in websites today');

```
insert into books (isbn, title, author1, author2, author3, category, subcategory,
quantity,  publisher, publish_date_edition, price, image, description) values('111-1-
1111-1131-9', 'Practical Web Services', 'B.M.Harwani','NULL', 'NULL','Computers &
Internet', 'Web Development', 100, 'Microchip Education','January 2010 : First Edition',
15.99,'images/111-1-1111-1131-9.jpg','jQuery has been around for a few years now and is
already pretty much the #1 or #2 JavaScript library used in websites today');

insert into books (isbn, title, author1, author2, author3, category, subcategory,
quantity,  publisher, publish_date_edition, price, image, description) values('111-1-
1111-1132-0', 'jQuery Recipes', 'B.M.Harwani','NULL', 'NULL','Computers & Internet', 'Web
Development', 100, 'Microchip Education','Feburary 2010 : First Edition',
19.99,'images/111-1-1111-1132-0.jpg','jQuery has been around for a few years now and is
already pretty much the #1 or #2 JavaScript library used in websites today');

insert into books (isbn, title, author1, author2, author3, category, subcategory,
quantity,  publisher, publish_date_edition, price, image, description) values('111-1-
1111-1132-1', 'Practical ASP.NET 3.5 Projects', 'B.M.Harwani','NULL', 'NULL','Computers &
Internet', 'Web Development', 100,  'Microchip Education','April 2010 : First Edition',
39.99,'images/111-1-1111-1132-1.jpg','jQuery has been around for a few years now and is
already pretty much the #1 or #2 JavaScript library used in websites today');

insert into books (isbn, title, author1, author2, author3, category, subcategory,
quantity,  publisher, publish_date_edition, price, image, description) values('111-1-
1111-1132-2', 'Practical EJB Project', 'B.M.Harwani','NULL', 'NULL','Computers &
Internet', 'Web Development', 100, 'Microchip Education','March 2010 : First Edition',
25.99,'images/111-1-1111-1132-2.jpg','jQuery has been around for a few years now and is
already pretty much the #1 or #2 JavaScript library used in websites today');

insert into books (isbn, title, author1, author2, author3, category, subcategory,
quantity,  publisher, publish_date_edition, price, image, description) values('111-1-
1111-1132-3', 'Foundation Joomla', 'B.M.Harwani','Bintu Mott', 'NULL','Computers &
Internet', 'Web Development', 100, 'Microchip Education','May 2010 : First Edition',
20.99,'images/111-1-1111-1132-3.jpg','jQuery has been around for a few years now and is
already pretty much the #1 or #2 JavaScript library used in websites today');

insert into books (isbn, title, author1, author2, author3, category, subcategory,
quantity,  publisher, publish_date_edition, price, image, description) values('111-1-
1111-1132-4', 'Web Development with AJAX', 'B.M.Harwani','Bintu Mott', 'David
Mortiz','Computers & Internet', 'Web Development', 100, 'Microchip Education','June 2010
: First Edition', 35.99,'images/111-1-1111-1132-4.jpg','jQuery has been around for a few
years now and is already pretty much the #1 or #2 JavaScript library used in websites
today');

insert into books (isbn, title, author1, author2, author3, category, subcategory,
quantity,  publisher, publish_date_edition, price, image, description) values('111-1-
1111-1132-5', 'Learn Mac OS X', 'B.M.Harwani','NULL', 'NULL','Computers & Internet',
'Operating Systems', 100, 'Microchip Education','March 2010 : First Edition',
25.99,'images/111-1-1111-1132-5.jpg','jQuery has been around for a few years now and is
already pretty much the #1 or #2 JavaScript library used in websites today');

insert into books (isbn, title, author1, author2, author3, category, subcategory,
quantity,  publisher, publish_date_edition, price, image, description) values('111-1-
1111-1132-6', 'Learn Adobe Practically', 'B.M.Harwani','Bintu Mott', 'NULL','Computers &
Internet', 'Graphic Design', 100, 'Microchip Education','May 2010 : First Edition',
20.99,'images/111-1-1111-1132-6.jpg','jQuery has been around for a few years now and is
already pretty much the #1 or #2 JavaScript library used in websites today');

insert into books (isbn, title, author1, author2, author3, category, subcategory,
quantity,  publisher, publish_date_edition, price, image, description) values('111-1-
1111-1132-7', 'Secure your Hardware & Network', 'B.M.Harwani','Bintu Mott', 'David
Mortiz','Computers & Internet', 'Security & Encryption', 100, 'Microchip Education','June
2010 : First Edition', 35.99,'images/111-1-1111-1132-7.jpg','jQuery has been around for a
```

few years now and is already pretty much the #1 or #2 JavaScript library used in websites today');

insert into books (isbn, title, author1, author2, author3, category, subcategory, quantity, publisher, publish_date_edition, price, image, description) values('111-1-1111-1132-8', 'Software Designing', 'B.M.Harwani','NULL', 'NULL','Computers & Internet', 'Software', 100, 'Microchip Education','March 2010 : First Edition', 25.99,'images/111-1-1111-1132-8.jpg','jQuery has been around for a few years now and is already pretty much the #1 or #2 JavaScript library used in websites today');

insert into books (isbn, title, author1, author2, author3, category, subcategory, quantity, publisher, publish_date_edition, price, image, description) values('111-1-1111-1132-9', 'Assemble your PC', 'B.M.Harwani','Bintu Mott', 'NULL','Computers & Internet', 'Hardware', 100, 'Microchip Education','May 2010 : First Edition', 20.99,'images/111-1-1111-1132-9.jpg','jQuery has been around for a few years now and is already pretty much the #1 or #2 JavaScript library used in websites today');

insert into books (isbn, title, author1, author2, author3, category, subcategory, quantity, publisher, publish_date_edition, price, image, description) values('111-1-1111-1133-0', 'iPhone SDK Programming Quickly & Easily', 'B.M.Harwani','Bintu Mott', 'David Mortiz','Computers & Internet', 'Mobile & Wireless Computing', 100, 'Microchip Education','June 2010 : First Edition', 35.99,'images/111-1-1111-1133-0.jpg','jQuery has been around for a few years now and is already pretty much the #1 or #2 JavaScript library used in websites today');

insert into books (isbn, title, author1, author2, author3, category, subcategory, quantity, publisher, publish_date_edition, price, image, description) values('111-1-1111-1141-0', 'Hollywood Gossips', 'B.M.Harwani','NULL', 'NULL','Entertainment', 'Movies', 100, 'Microchip Education','January 2010 : First Edition', 15.99,'images/111-1-1111-1141-0.jpg','Trying to find healthy and delicious recipes? Food Network makes that easy with their collection of low fat, low calorie and low carb recipes');

insert into books (isbn, title, author1, author2, author3, category, subcategory, quantity, publisher, publish_date_edition, price, image, description) values('111-1-1111-1141-1', 'Craft for Money', 'B.M.Harwani','NULL', 'NULL','Creative Crafts', 'Embroidery', 100, 'Microchip Education','Feburary 2010 : First Edition', 19.99,'images/111-1-1111-1141-1.jpg','Trying to find healthy and delicious recipes? Food Network makes that easy with their collection of low fat, low calorie and low carb recipes');

insert into books (isbn, title, author1, author2, author3, category, subcategory, quantity, publisher, publish_date_edition, price, image, description) values('111-1-1111-1141-2', 'Food & Exercises', 'B.M.Harwani','NULL', 'NULL','Health Related', 'Food', 100, 'Microchip Education','April 2010 : First Edition', 39.99,'images/111-1-1111-1141-2.jpg','Trying to find healthy and delicious recipes? Food Network makes that easy with their collection of low fat, low calorie and low carb recipes');

insert into books (isbn, title, author1, author2, author3, category, subcategory, quantity, publisher, publish_date_edition, price, image, description) values('111-1-1111-1141-3', 'Scream III', 'B.M.Harwani','NULL', 'NULL','Mysterious Stories', 'Horror', 100, 'Microchip Education','March 2010 : First Edition', 25.99,'images/111-1-1111-1141-3.jpg','Trying to find healthy and delicious recipes? Food Network makes that easy with their collection of low fat, low calorie and low carb recipes');

insert into books (isbn, title, author1, author2, author3, category, subcategory, quantity, publisher, publish_date_edition, price, image, description) values('111-1-1111-1141-4', 'Hong Kong Tourism Guide', 'B.M.Harwani','Bintu Mott', 'NULL','Travelling Guides', 'Places', 100, 'Microchip Education','May 2010 : First Edition', 20.99,'images/111-1-1111-1141-4.jpg','Trying to find healthy and delicious recipes? Food Network makes that easy with their collection of low fat, low calorie and low carb recipes');

Index

W

X